ScienceWorld 9

Australian Curriculum edition

Lesley Englert · Peter Stannard · Ken Williamson
BA, DipT · BSc, DipEd · BSc (Hons), DipEd

This edition published in 2021 by

Matilda Education Australia, an imprint
of Meanwhile Education Pty Ltd
Melbourne, Australia
T: 1300 277 235
E: customersupport@matildaed.com.au
www.matildaeducation.com.au

First edition published in 2012 by Macmillan Science and Education Australia Pty Ltd

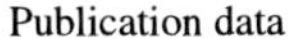

Publication data

Author: Lesley Englert, Peter Stannard and Ken Williamson

Title: ScienceWorld 9 Workbook: Australian Curriculum edition
ISBN: 978 1 4202 2996 7

Publisher: Peter Saffin
Project editor: Hannah Cartmel
Illustrator: Chris Dent, Brent Hagen, Guy Holt, Andy Craig and Nives Porcellato
Cover designer: Dim Frangoulis
Text designer: Megan Strange
Production control: Loran McDougall
Typeset in Helvetica, Trade Gothic, Segoe Print and Linotype Kaliber by Firefly Education Pty Ltd
Cover image: Photolibrary/Jim Reed (main photo), Shutterstock/Tischenko Irina (background detail)

Printed in Australia by Pegasus Media & Logistics
1 2 3 4 5 6 7 25 24 23

Internet addresses
At the time of printing, the internet addresses appearing in this book were correct. Owing to the dynamic nature of the internet, however, we cannot guarantee that all these addresses will remain correct.

Warning: It is recommended that Aboriginal and Torres Strait Islander peoples exercise caution when viewing this publication as it may contain images of deceased persons.

About this book

This workbook is full of exercises that will help you understand the material in the textbook, think about the information, relate it to your own experiences and write about it in your own words.

The exercises ask questions and pose problems to give you the opportunity to reflect on your knowledge, then show what you know. For example, in the exercise below you have to complete the cartoon.

Do you find that you know the answer but often do not know how to write it in the correct form? The workbook teaches you how to organise this information into many different written and oral forms.

On the following page you will see some sample pages from the workbook with descriptions of its various features. These descriptions will help you understand how to use your workbook.

Getting to know your workbook

Each chapter begins with the same chapter number and name as found in your textbook.

Sometimes you will see a Challenge box. These boxes contain challenging questions or problems that you might like to attempt.

This is called the Overview—it is a diagram of the way the ideas in the chapter are connected.

The section of the workbook called During reading contains exercises that ask you questions about the ideas on certain pages of your textbook.

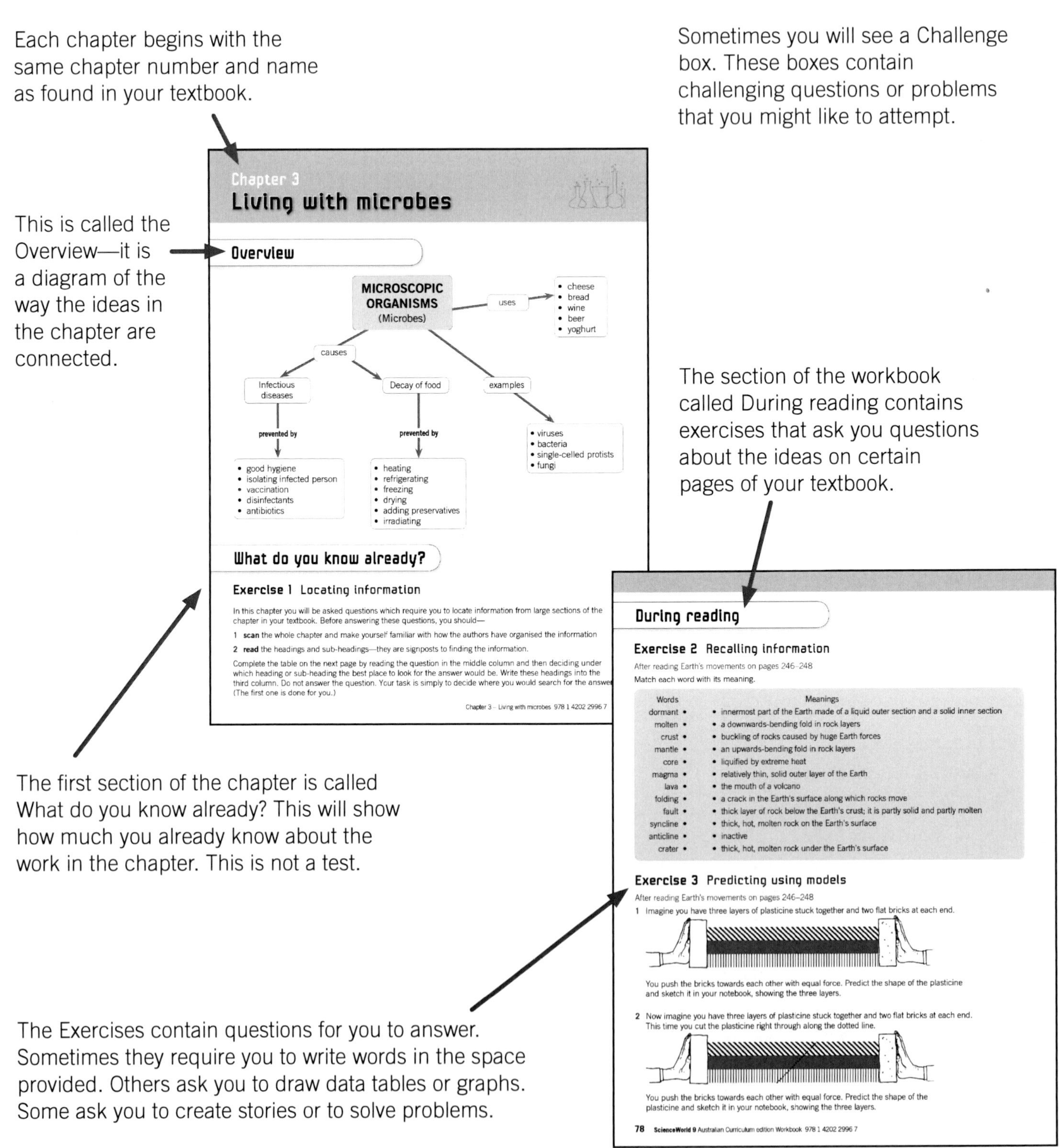

Chapter 3
Living with microbes

Overview

What do you know already?

Exercise 1 Locating information

In this chapter you will be asked questions which require you to locate information from large sections of the chapter in your textbook. Before answering these questions, you should—

1 **scan** the whole chapter and make yourself familiar with how the authors have organised the information

2 **read** the headings and sub-headings—they are signposts to finding the information.

Complete the table on the next page by reading the question in the middle column and then deciding under which heading or sub-heading the best place to look for the answer would be. Write these headings into the third column. Do not answer the question. Your task is simply to decide where you would search for the answe[r] (The first one is done for you.)

Chapter 3 – Living with microbes 978 1 4202 2996 7

During reading

Exercise 2 Recalling information

After reading Earth's movements on pages 246–248

Match each word with its meaning.

Words	Meanings
dormant •	• innermost part of the Earth made of a liquid outer section and a solid inner section
molten •	• a downwards-bending fold in rock layers
crust •	• buckling of rocks caused by huge Earth forces
mantle •	• an upwards-bending fold in rock layers
core •	• liquified by extreme heat
magma •	• relatively thin, solid outer layer of the Earth
lava •	• the mouth of a volcano
folding •	• a crack in the Earth's surface along which rocks move
fault •	• thick layer of rock below the Earth's crust; it is partly solid and partly molten
syncline •	• thick, hot, molten rock on the Earth's surface
anticline •	• inactive
crater •	• thick, hot, molten rock under the Earth's surface

Exercise 3 Predicting using models

After reading Earth's movements on pages 246–248

1 Imagine you have three layers of plasticine stuck together and two flat bricks at each end.

You push the bricks towards each other with equal force. Predict the shape of the plasticine and sketch it in your notebook, showing the three layers.

2 Now imagine you have three layers of plasticine stuck together and two flat bricks at each end. This time you cut the plasticine right through along the dotted line.

You push the bricks towards each other with equal force. Predict the shape of the plasticine and sketch it in your notebook, showing the three layers.

78 ScienceWorld 9 Australian Curriculum edition Workbook 978 1 4202 2996 7

The first section of the chapter is called What do you know already? This will show how much you already know about the work in the chapter. This is not a test.

The Exercises contain questions for you to answer. Sometimes they require you to write words in the space provided. Others ask you to draw data tables or graphs. Some ask you to create stories or to solve problems.

Roundup will check how much you know about the main ideas in the chapter. Be prepared to write answers, teach someone else, have a group discussion or role play.

Chapter 1

Science is investigating

Overview

1 Planning the experiment

2 Conducting the experiment

3 Processing data

4 Evaluating the experiment

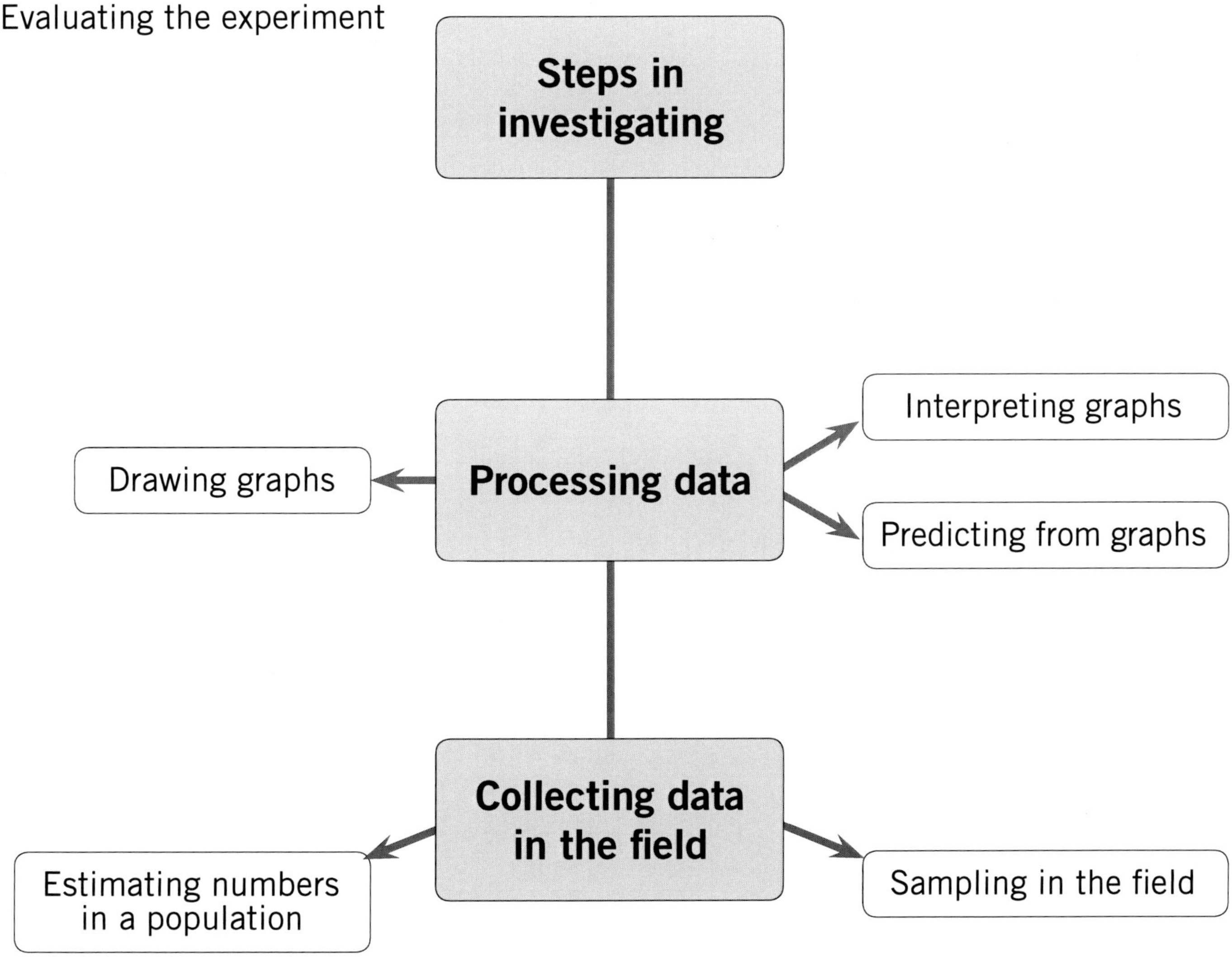

Steps in investigating

Processing data

Drawing graphs

Interpreting graphs

Predicting from graphs

Collecting data in the field

Estimating numbers in a population

Sampling in the field

During reading

Exercise 1 Critical reading

After reading page 3

1 List the four steps in investigating. ______

2 List three things you can do to the variables in an experiment. ______

3 In which step would you interpret the data you have collected? ______

4 In Step 4, how will you know whether or not you need to modify your hypothesis? ______

5 Why is there an arrow from Step 4 back to Step 1? ______

Exercise 2 Identifying variables

After doing the Activity on page 4

Anthony is designing an experiment to investigate whether it is true that people with low blood pressure are not bitten by mosquitoes as often as people with higher blood pressure.

1 Which is the independent variable in Anthony's experiment?

______

2 Which is the dependent variable? ______

3 Suggest ways of measuring the dependent variable. ______

4 List the variables Anthony will need to control. ______

Exercise 3 Evaluating

After reading Evaluating an experiment on page 4

Lauren and Amy worked separately to solve this problem.

A cube of sugar has about the same mass as a level teaspoon of sugar crystals. Which will dissolve more quickly?

LAUREN

Method

1 Fill two glasses with tap water.

2 At the same time add a teaspoon of sugar crystals to one and a cube of sugar to the other.

Results

The cube broke up into a pile of crystals and took about the same time to dissolve as the teaspoon of crystals.

Conclusion

1 Write a conclusion for Lauren's experiment.

____

____

____

AMY

Method

1 Fill a glass with water from the tap.

2 Add a teaspoon of sugar crystals and stir.

3 Use a stopwatch to measure how long it takes for all the sugar to dissolve.

4 Do this three times.

5 Repeat the experiment using a sugar cube in a glass of tap water.

Results

	Time to dissolve			
	Trial 1	Trial 2	Trial 3	Average
Sugar crystals	69	65	61	65
Cube of sugar	66	74	76	72

Conclusion

2 Write a conclusion for Amy's experiment.

____

____

____

3 Lauren claimed that her experiment was better because she solved the problem more quickly than Amy did. Do you agree with her? Explain. ______

______

4 How can you explain the fact that Amy obtained different results in her three trials? ______

______

______

5 Why did Amy calculate the average in each case? ______

______

6 Would Amy's conclusion have been different if she had not repeated her measurements? Explain.

______

______

______

Exercise 4 Graphing

After doing Experiment 1 on page 7

Tatiana used a newton spring balance to measure the force needed to pull two long pieces of Velcro apart as shown. The Velcro strips were 4 cm wide and 10 cm long. She varied the length of Velcro that was in contact. Her results are shown in the table below.

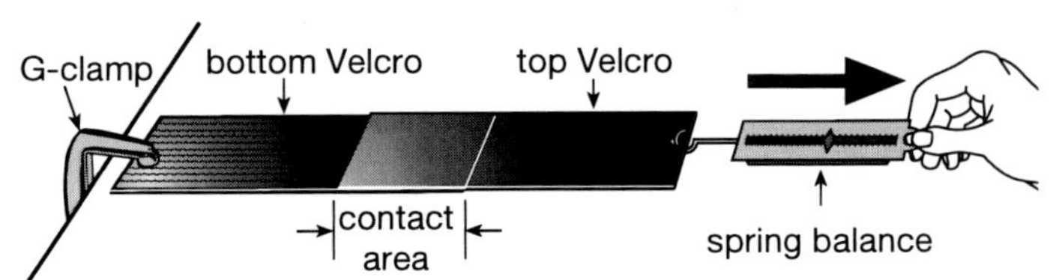

Length of Velcro in contact	Force needed (newtons)			
	Trial 1	Trial 2	Trial 3	Average
1 cm	38	38	40	______
2 cm	86	80	78	______
3 cm	130	126	126	______
4 cm	149	166	165	______
5 cm	218	198	199	______
6 cm	241	216	248	______
7 cm	291	281	277	______
8 cm	318	315	326	______

1 Which is the independent variable in Tatiana's experiment? ______________________

2 Which is the dependent variable? ______________________

3 For each length of Velcro, calculate the average force needed. Add these to Tatiana's data table.

4 Use Tatiana's data to draw a best-fit graph. (Use a sheet of graph paper.)

5 Write a generalisation describing the relationship between the two variables.

____

____

6 Complete the following sentences.

An increase in the length of Velcro in contact causes an increase in ______________________

______________________ . Similarly, a decrease in the length of Velcro causes a ______________________

in ______________________ . The fact that the graph is a straight line means that the

increases or decreases are ______________________ equal. For example, if you double the

length of Velcro you ______________________ the force needed, and if you triple the

length you ______________________ .

7 Use your graph to make these predictions:

(a) How much force is needed to pull the Velcro apart if there is 4.5 cm in contact? ______________________

(b) How much Velcro is in contact if the force needed is 300 newtons? ______________________

(c) How much force is needed if there is 10 cm of Velcro in contact? ______________________

Exercise 5 Understanding on three levels

After reading Experiments using animals on pages 8–9

Level 1: Reading for accuracy (Be able to show exactly where statements in the textbook support your answer.)

1 What was the name of the drug found to produce deformities in newborn babies? ______________________

2 Which animals are usually used in laboratory tests of new drugs? ______________________

____

3 What is a placebo? ______________________

____

Level 2: Drawing conclusions (Be able to use information in the textbook to show how you arrived at your conclusion.)

1 What is the difference between the test group and the control group in an experiment?

____

2 Why is it sometimes necessary to use a blind experimental design? ______________________

____

Level 3: Applying your knowledge (Be able to apply your knowledge in a wider context.)

1 Give an example of an investigation where it could be important to use a double-blind experimental design.

____

____

2 The manufacturers of a new drug claim that it can help smokers give up smoking. A blind experimental design is used to test the drug, and 10% of the test group give up smoking while 5% of the control group give up smoking. How would you interpret this result?

____

____

____

3 The use of the drug thalidomide has been banned. However, thalidomide has recently been found to be useful in the treatment of rheumatoid arthritis, leprosy and even HIV (AIDS). What would need to be done before the ban on thalidomide could be lifted?

____

____

Exercise 6 Interpreting graphs

After reading pages 14–15

Complete the sentences describing each of the three graphs.

Graph 1 There is a high ______________ between the two variables. The taller the father, the ______________ his son is likely to be. There is a ______________ relationship between the height of the father and the height of his son.

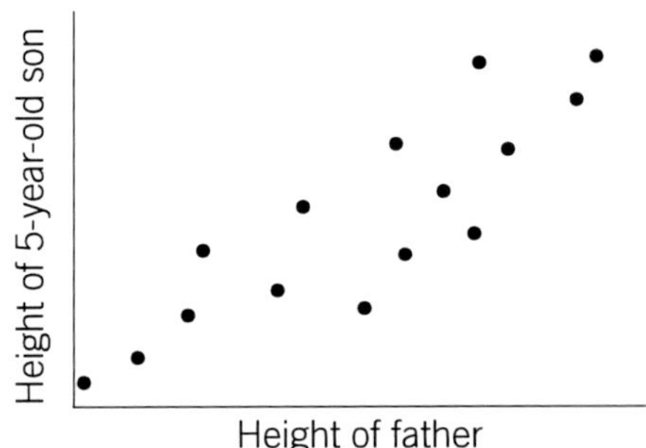

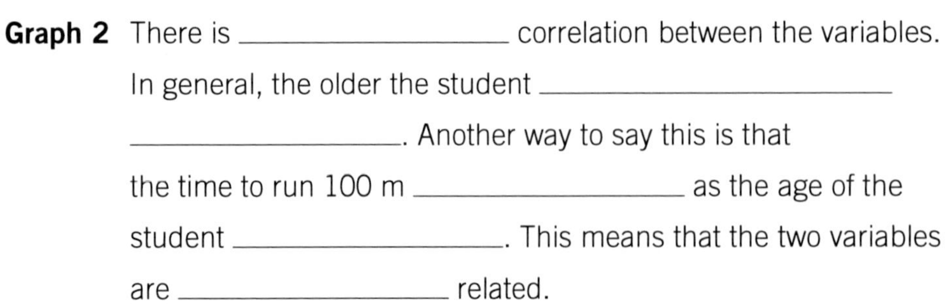

Graph 2 There is ______________ correlation between the variables. In general, the older the student ______________ ______________. Another way to say this is that the time to run 100 m ______________ as the age of the student ______________. This means that the two variables are ______________ related.

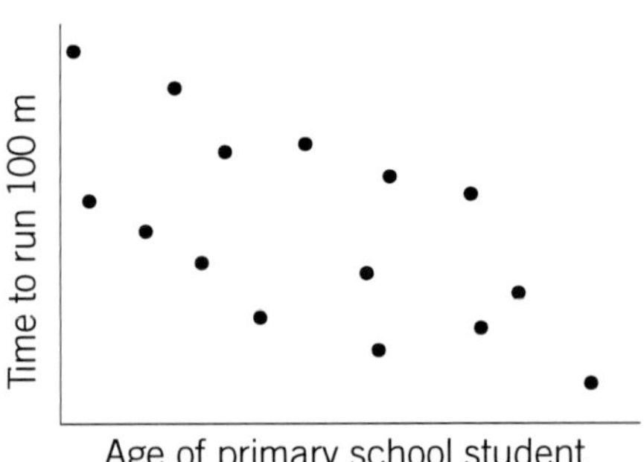

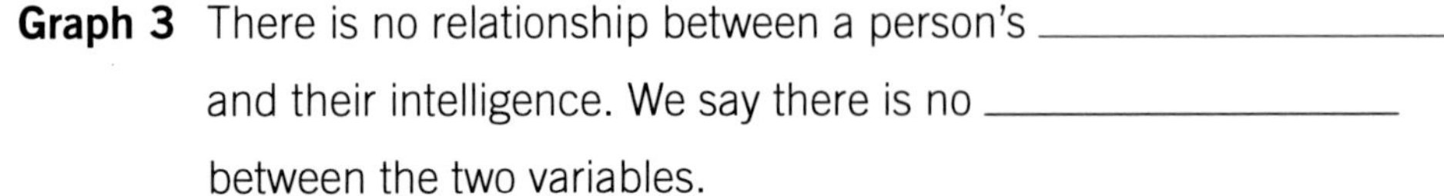

Graph 3 There is no relationship between a person's ______________ and their intelligence. We say there is no ______________ between the two variables.

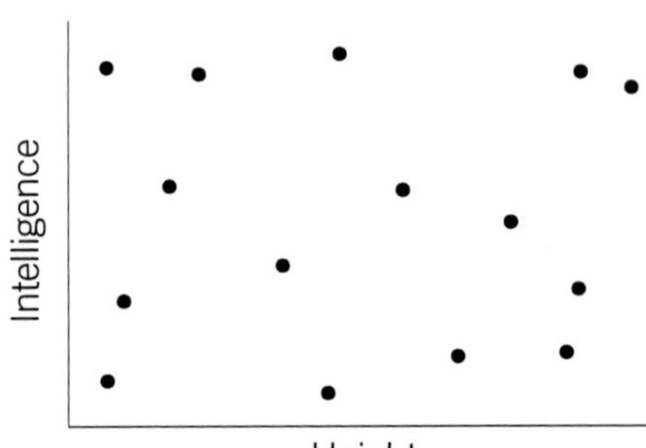

Roundup

Exercise 7 Planning an experiment

Jacob and his mates often go camping. One hot night he notices that the crickets are particularly noisy. This causes him to wonder whether the chirp rate of crickets depends on the temperature.

He decides to investigate this. How should he go about it?

Use the information on page 3 of your textbook to plan an experiment that could solve Jacob's problem. Write your plan in the working space below using the four steps and the pointers under the steps.

Working space

Chapter 2
Light and sound

Overview

Light and sound

- both are forms of energy
- both can be converted into forms of energy
- both are reflected (light as an image and sound as an echo) by other objects

Sound

- produced as sound waves by vibrating objects
- travels in all directions through solids, liquids and gases, moving from particle to particle
- travels at 330 m/s

Light

- produced as electromagnetic radiation waves
- can travel through a vacuum and transparent substances in a straight line
- travels at 3 000 000 m/s
- can be refracted (bent) when it passes from one transparent substance to another
- can be dispersed (split by a prism into the seven colours of the spectrum)

What do you know already?

Exercise 1 Working out and applying rules

1 Tranh noticed that when he saw **5** in the mirror, it looked like **Ƨ**, while **F** looked like **ꟻ**, and **➜** looked like **⬅**.
Work out a rule that tells you what happens to a reflected image.

2 Use the rule to find the word the mirror image of this word puzzle would be. **8ɘnut4nu**

3 A mirror is attached to one side of a protractor and a ray of light shone onto it.
The paths taken by the incoming ray and the reflected ray are shown in the three diagrams below.

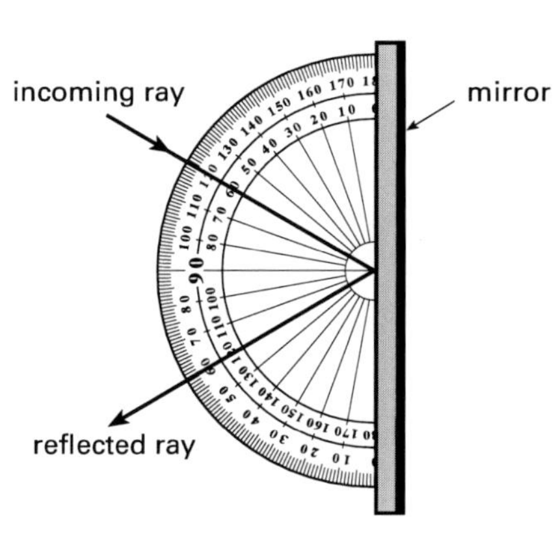

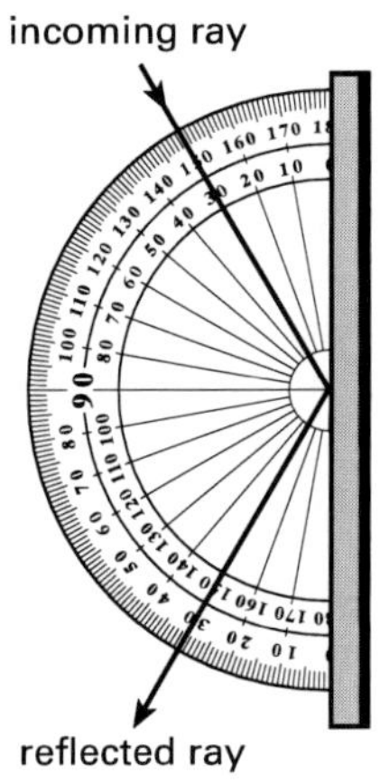

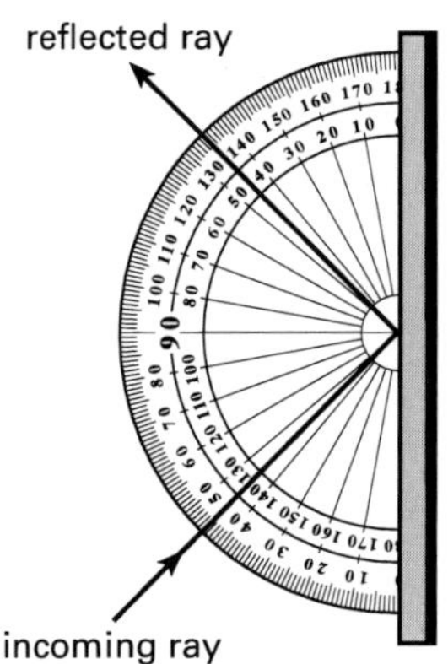

Work out a rule to explain these observations.________________________________

____

4 Use your rule to write answers to these questions.
At which angle will an incoming light ray reflect from a mirror if it hits the mirror at a—

48° angle? ________________ 74° angle? ________________

Exercise 2 Observing

Find four mistakes that the artist has made in this cartoon.

1 ____

____

2 ____

____

3 ____

____

4 ____

____

During reading

Exercise 3 Using comparison linking words

After reading Properties of light and sound on page 29

Light and sound have several similar characteristics.
The *Comparison linking words* on page 92 of your workbook are used in sentences to join similar information.

Join the sentences below into one sentence using a different *Comparison linking word* each time.
The first pair is done for you as an example.

1 Light is a form of energy.
Sound is a form of energy.
Light and sound are *both* forms of energy.

2 Light can be reflected from objects.
Sound can be reflected from objects.

3 Light can be converted to other forms of energy.
Sound can be converted to other forms of energy.

Exercise 4 Completing and measuring diagrams

After reading The law of reflection on pages 30–31

1 For each diagram, use lines and arrows to show the direction in which the rays of light bounce after hitting the mirror.

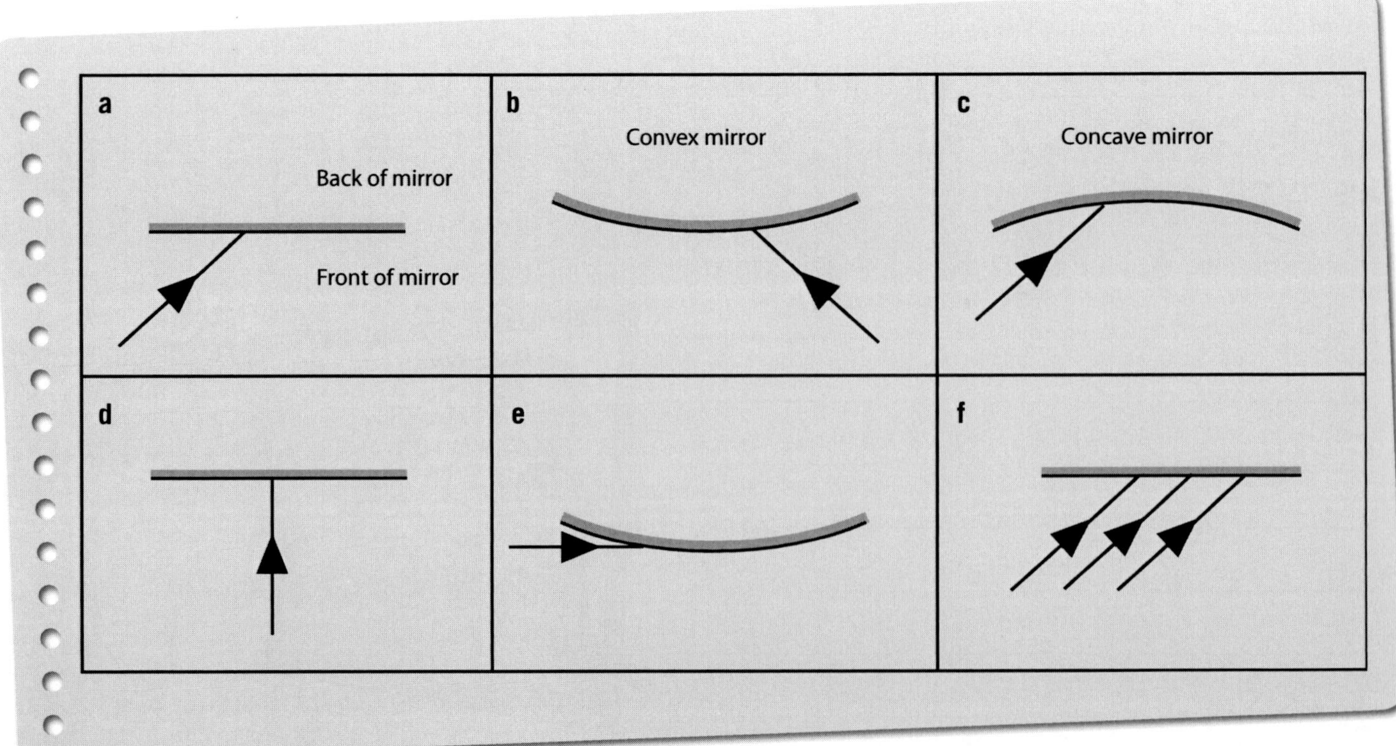

2 Submarines use periscopes while under water to view objects above the water. Because light travels only in straight lines, mirrors are needed to reflect the light from the surface of the water to the eye looking in the periscope.

In the diagram on the right draw two mirrors in the correct positions to make the periscope work. Also draw the path of the light through the periscope.

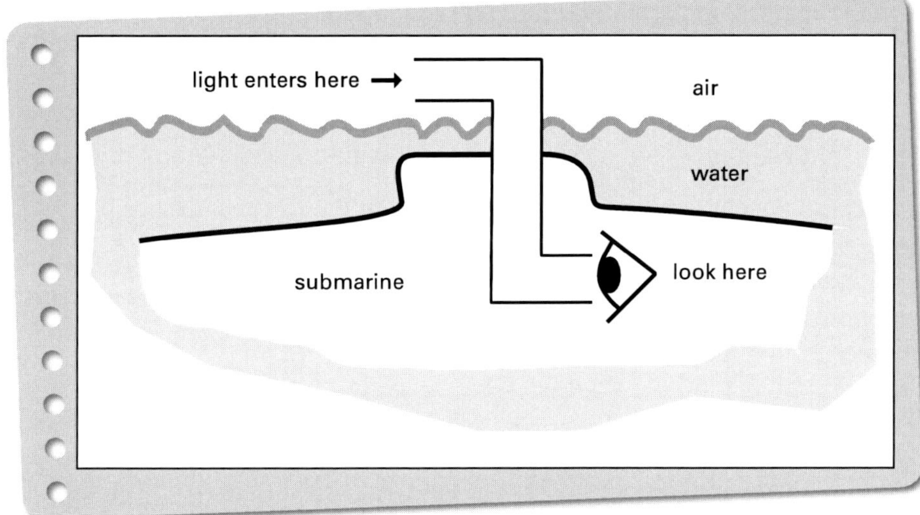

3 In the following diagrams, measure the focal length of the mirror in millimetres.

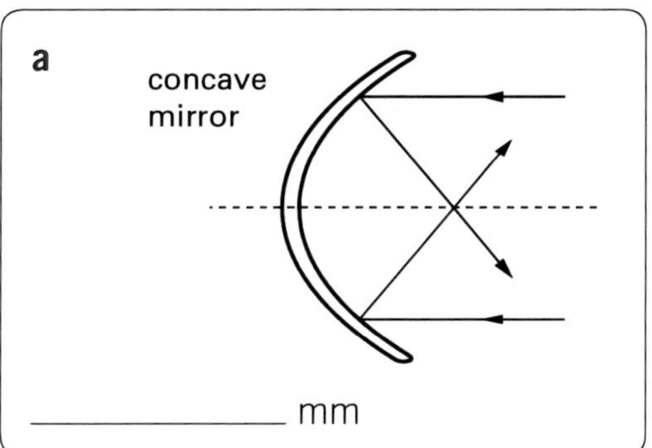

_______________ mm

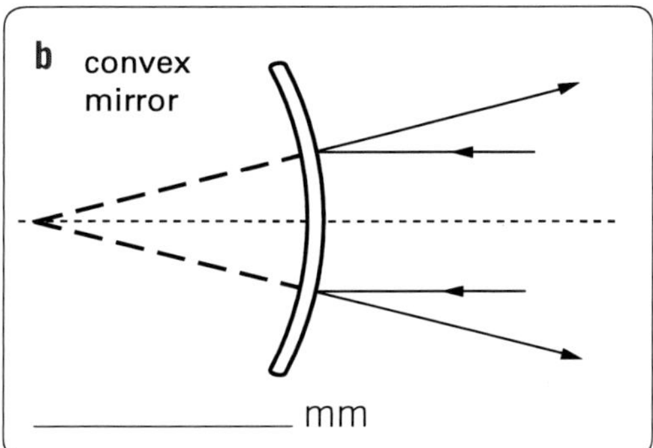

_______________ mm

Exercise 5 Generalising

After reading Refraction of light on pages 32–33

1 The bending of light makes it difficult to judge the depth of objects under water. The diagram shows the actual position of the fish. Sketch where the fish would appear to be to the observer.

2 Complete this generalisation.

When viewed at an angle from above the surface of the water, objects under the water are usually _______________________ than they appear to be.

Exercise 6 Predicting

After reading pages 32–33 and doing Investigation 5 on pages 33–34

Predict and sketch what will happen to the three light rays as they pass through this series of lenses. (Hint: show the rays bending in the centre of each lens.)

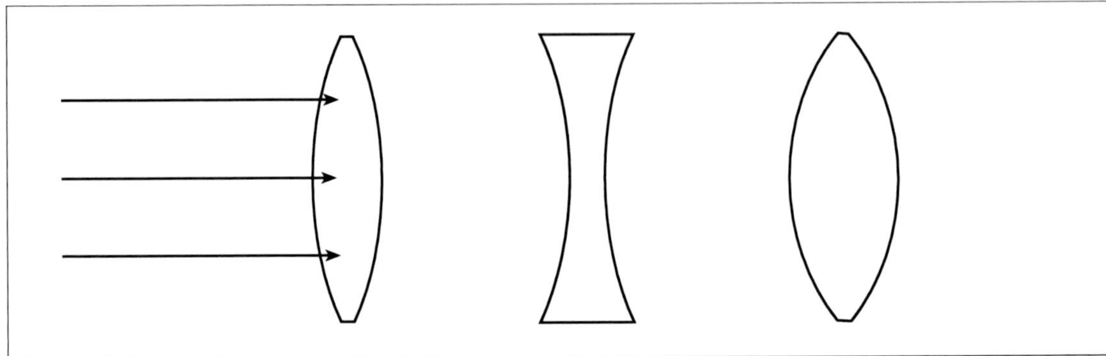

Exercise 7 Labelling

After reading How the eye focuses light on page 36

When you are asked to label a diagram, you should indicate the parts as clearly as possible so that there is no mistake about which part the label refers to. You should—

- use a ruler to rule a straight line with an arrow joining the label to the part
- make sure that you do not have any lines crossing other lines
- print the label neatly.

1 Study the illustration of the eye on page 36 (Fig 6) of your textbook. Close your textbook and label the diagram opposite with the following:

lens	top eyelid
iris	lower eyelid
retina	lens muscle
pupil	optic nerve
cornea	jelly-like substance
fovea	

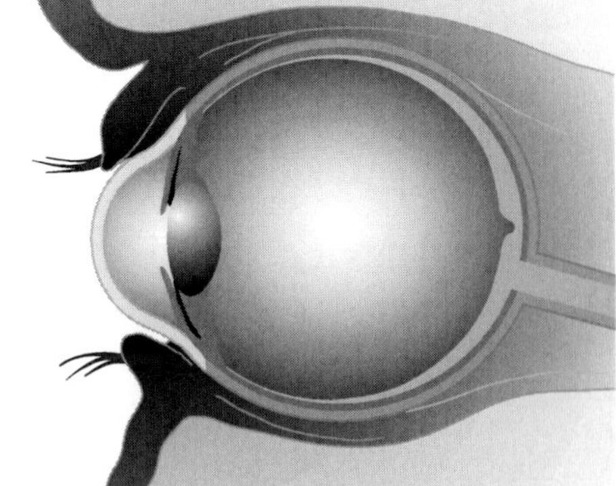

2 When you have checked that the labels are correct, add the name against each description below.

- clear, thin 'skin' that covers the front of the eye —
- coloured ring of muscle which controls the amount of light entering the eye —
- the opening in the iris through which light passes —
- bundle of nerves which carry messages to the brain —
- the back of the eyeball which contains vision receptors —
- part which focuses light onto the retina —

3 What can the lens in your eye do that glass lenses cannot? (Use a *Contrast linking word* from page 92 in your answer.)

Exercise 8 Applying information

After reading Light and colour on page 39

When light enters a prism, it is dispersed into the colours of the spectrum because each colour has its own wavelength, and each wavelength is bent slightly differently as it passes from air to glass. Violet light is bent most, and red is bent least.

Suppose a ray of white light enters a large glass ball as shown on the right.

Write the correct colour for each of the seven rays coming out of the glass ball.

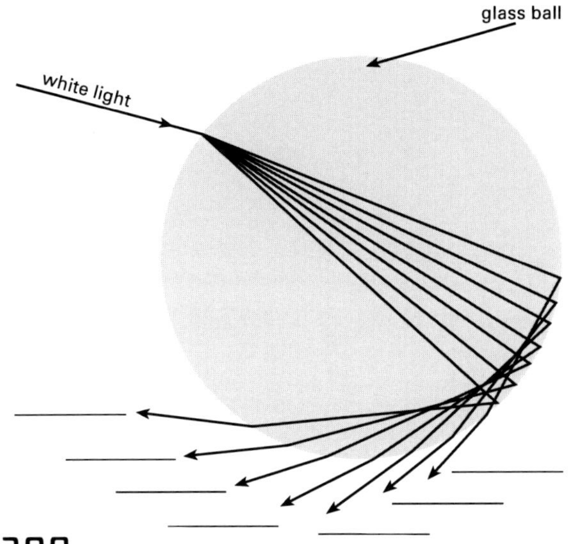

Exercise 9 Explaining in simple language

After reading Why are things coloured? on page 40

Imagine a six-year-old asks you the following questions: Why is butter yellow? Why isn't it red or blue? How would you reply without using too much scientific language so that the child understands about reflected and absorbed light?

____

____

____

____

Exercise 10 Synthesising

After reading Making colours on page 42

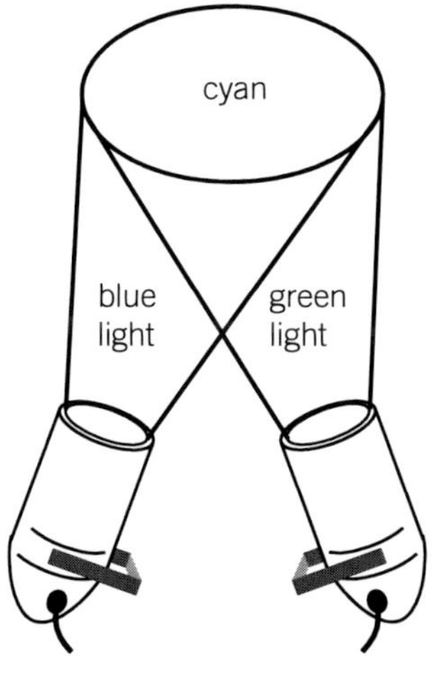

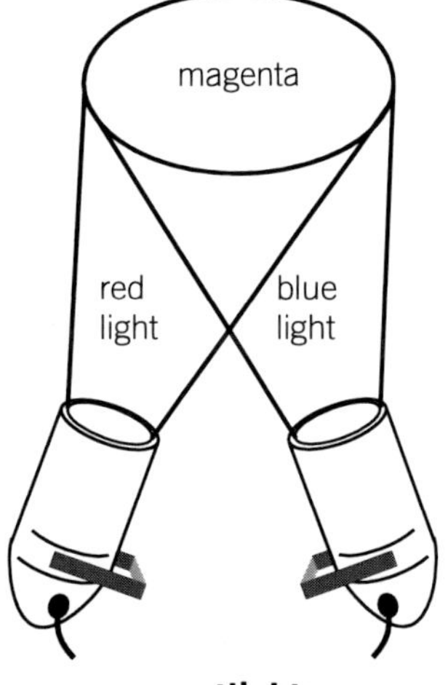

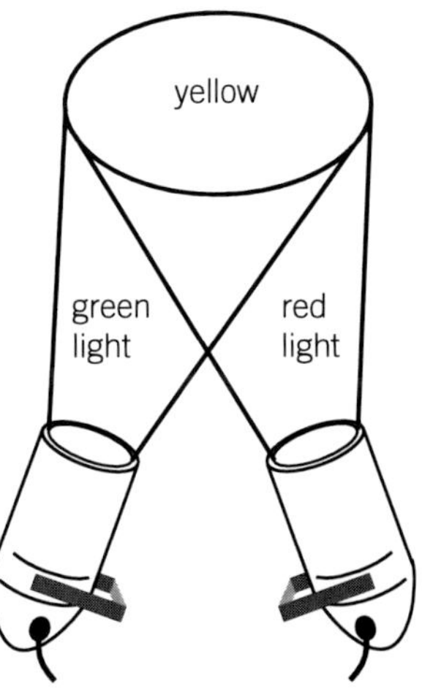

spotlights

When using coloured spotlights:

- blue, red and green are called primary colours.
- primary colours cannot be made by mixing other colours.
- when all three primary colours are mixed they produce white light.

Use the information on the previous page to work out the colour for each of the three overlapping areas in the diagram below.

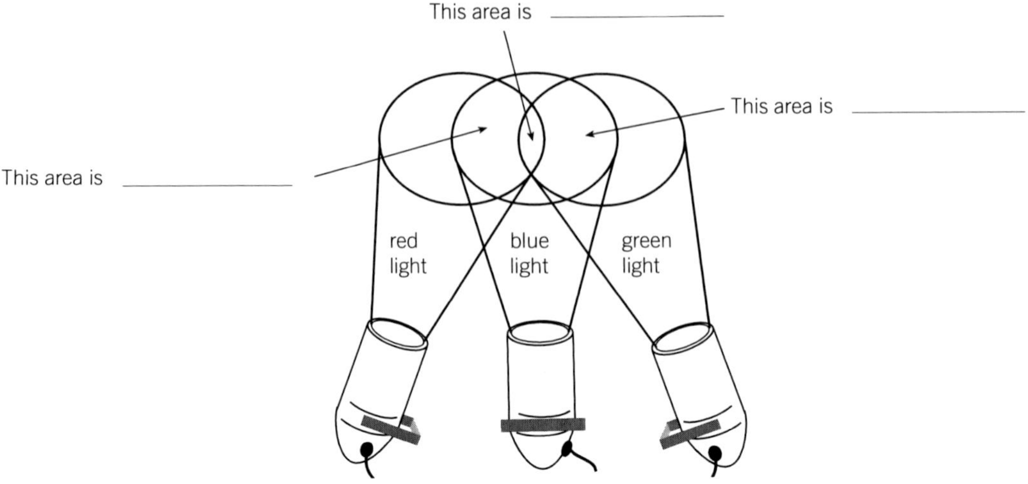

Exercise 11 Extrapolating

After reading Seeing colours on page 43

1 If the cone cells in the retina are only sensitive to blue light, green light and red light, how do we see yellow light?

____

____

2 If the seven colours of the spectrum make up white light, why do we get a black mess when we mix seven pots of paint of these colours?

____

____

Exercise 12 Interpreting data

After reading Light and sound waves on pages 47–48

Use the table on page 48 of your textbook to decide whether these statements are true or false. Circle your decision and be ready to justify your answer.

	Statement		
1	You will hear a friend's voice yelling across an oval sooner on a cold day than on a hot day.	TRUE	FALSE
2	You will hear a noise underwater sooner than if you heard it from the same distance above the water.	TRUE	FALSE
3	At 20°C, sound travels faster through wood than through granite.	TRUE	FALSE
4	At 0°C, sound travels faster through oxygen than through air.	TRUE	FALSE
5	Iron is a better conductor of sound than copper.	TRUE	FALSE

Exercise 13 Answering multiple-choice questions

After reading Light waves on page 49 and Light waves and refraction on page 51

Below are some multiple-choice questions. In this type of question, you usually have four possible answers. One will be correct and the others will be incorrect or only partly correct.

Read the incomplete sentence and the four possible answers very carefully. Then circle the letter that you think is correct to complete the sentence. The first one is done for you.

1 Ultraviolet light has a longer wavelength than—

(a) microwaves ⟶ incorrect

(b) X-rays and gamma rays ⟶ correct

(c) infra-red radiation ⟶ incorrect

(d) visible light ⟶ incorrect

2 Radiation from the sun comes to us in the form of—

(a) light waves

(b) microwaves

(c) X-rays

(d) all of the above

3 You hear thunder after you see lightning because—

(a) your eyes are more sensitive than your ears.

(b) your brain recognises light before it recognises sound.

(c) light travels through air about a million times faster than sound does.

(d) the lightning actually strikes several seconds before the thunderclap occurs.

4 A rainbow forms when sunlight passes through raindrops because—

(a) each of the colours which make up white light has its own wavelength.

(b) raindrops act as a prism to refract the white light.

(c) the raindrops disperse the different wavelengths to form a rainbow.

(d) all of the above.

5 As a ray of light travelling through air passes through glass—

(a) its speed decreases.

(b) its speed increases.

(c) its speed does not change.

(d) its speed decreases as it hits the glass, then increases as it travels through the glass.

Roundup

Exercise 14 Writing multiple-choice questions

In the previous exercise, you answered several multiple-choice questions. Now it's your turn to create your own multiple-choice answers to try out on your classmates.

Remember, a multiple-choice question has two parts:

- the stem, which may be a question, for example, How does light travel? or an unfinished statement, for example, Light can travel—
- usually four possible answers, one of which is correct and the other three which are incorrect or only partly correct (but not so 'way out' that the correct answer is too easy to find!).

Below are three unfinished statements. For each statement add four possible answers, only one of which is correct.

1 Both light and sound—

a ______________________

b ______________________

c ______________________

d ______________________

2 Light is refracted when—

a ______________________

b ______________________

c ______________________

d ______________________

3 Dispersion is the process in which—

a ______________________

b ______________________

c ______________________

d ______________________

4 Sound waves are produced by—

a ______________________

b ______________________

c ______________________

d ______________________

Chapter 3

Living with microbes

Overview

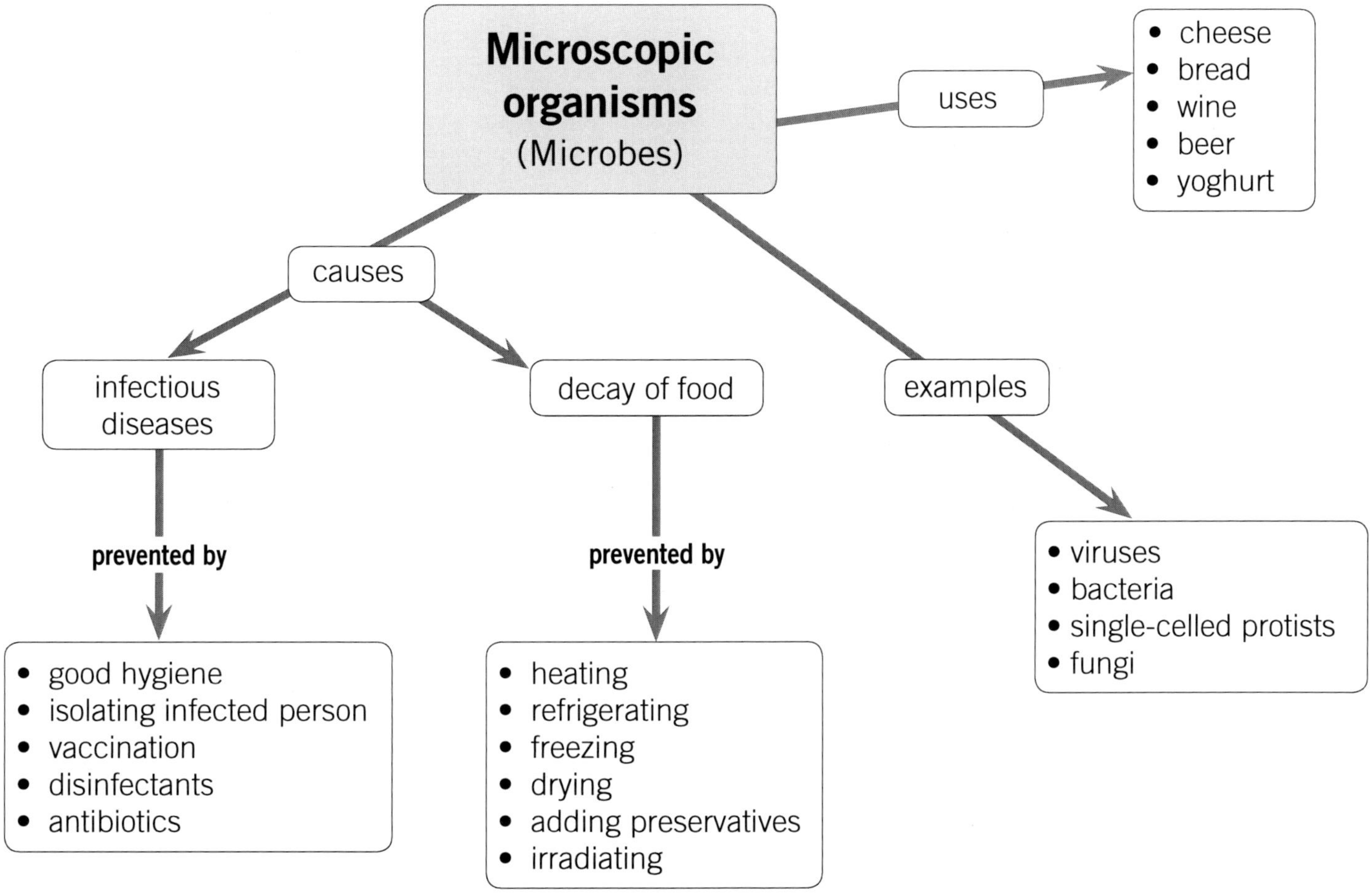

Microscopic organisms (Microbes)

- uses → cheese, bread, wine, beer, yoghurt
- causes → infectious diseases → prevented by: good hygiene, isolating infected person, vaccination, disinfectants, antibiotics
- causes → decay of food → prevented by: heating, refrigerating, freezing, drying, adding preservatives, irradiating
- examples → viruses, bacteria, single-celled protists, fungi

What do you know already?

Exercise 1 Locating information

In this chapter you will be asked questions which require you to locate information from large sections of the chapter in your textbook. Before answering these questions, you should—

1 **scan** the whole chapter and make yourself familiar with how the authors have organised the information

2 **read** the headings and sub-headings—they are signposts to finding the information.

Complete the table on the next page by reading the question in the middle column and then deciding under which heading or sub-heading the best place to look for the answer would be. Write these headings into the third column. Do not answer the question. Your task is simply to decide where you would search for the answer. (The first one is done for you.)

Main headings and sub-headings used in Chapter 4	Questions	I would expect to find the answer in ...
3.1 Microscopic life • Protists • Micro-organisms in pond water • Bacteria **3.2 Helpful microbes** • Decomposers • Making foods • Microbes in the gut • Stopping the decay of food **3.3 Microbes and disease** • Causes of disease • Fighting disease • Controlling infections	**1** How do protists move about? **2** List six ways in which vegetables can be preserved. **3** How can cholera be prevented? **4** Which microbes are used to make ginger beer? **5** Which product of baking bread may be harming the ozone layer? **6** What are the two types of white blood cells that destroy the measles virus? **7** True or false? Fungi are a type of bacteria.	**Protists**

Exercise 2 Understanding the information

When you have located the section containing the information that you need, you should read it carefully to understand how the information is organised.

To the right is a paragraph from the textbook.

Underlining has been added.

Read it through carefully and then answer the questions below.

> Many protists have structures on the outside of the cell to help them move through the water. For example, some have tiny hair-like *cilia* (SILL-ee-a) on the outside of the cell. These beat rhythmically and propel the organism through the water. Others have large whip-like *flagella* (fla-JELL-a) which act like paddles to move the organism.

1 What is the main idea of this paragraph? ______________________

2 What does them refer to? ______________________

3 What does some refer to? ______________________

4 In (SILL-ee-a) and (fla-GELL-a) why are some letters in CAPITALS and some letters in lower case?

____

5 What does These refer to? ______________________

6 What does Others refer to? ______________________

Exercise 3 Understanding diagrams

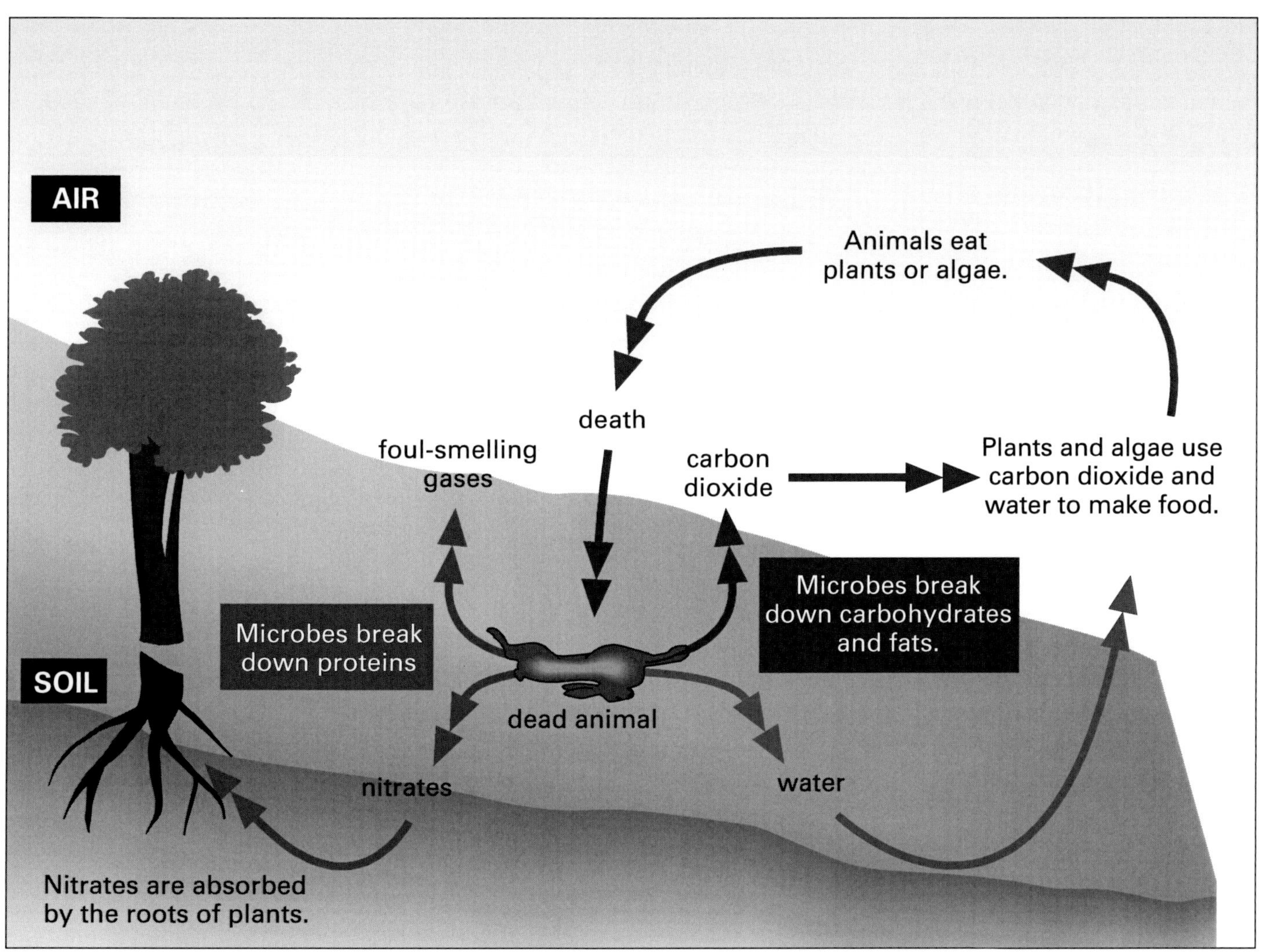

Authors often use diagrams to illustrate the text. Most diagrams use basic sketches, with labels and arrows to illustrate a process. Examine the diagram above carefully and then answer the following questions.

1 When an animal dies, what are the four substances released? ____________________

____

2 What happens to the carbon dioxide? ____________________

____

3 What happens to the nitrates? ____________________

____

4 Which product is the only substance not recycled? ____________________

____

5 List the three substances which are broken down by microbes ____________________

____

Exercise 4 Recalling information

After reading the whole chapter

Follow the directions on the path and answer the questions on a separate sheet of paper. Do not continue on your journey after each STOP sign unless all your answers are correct or you know why your answers were incorrect. Answers are on page 22 of this workbook.

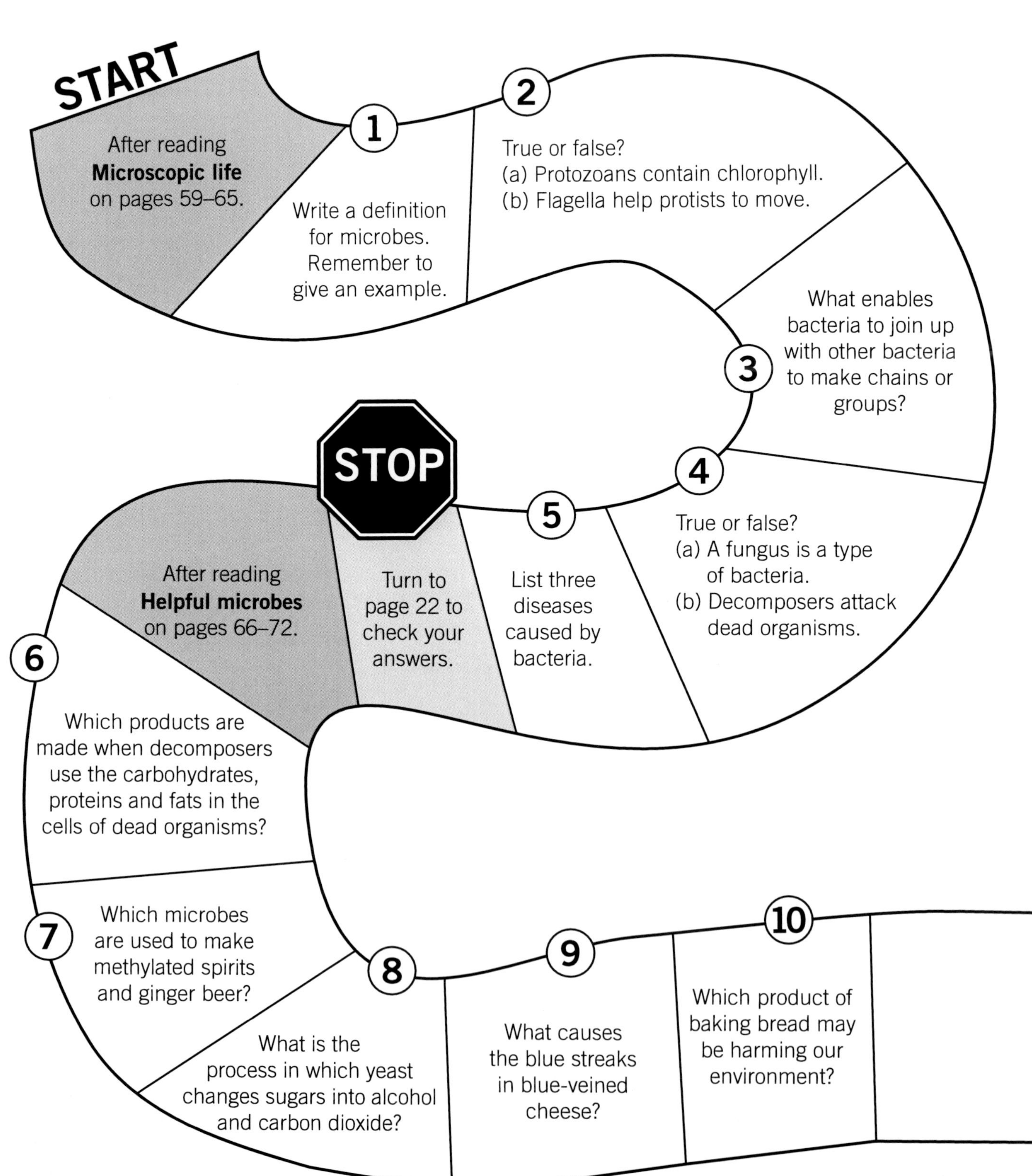

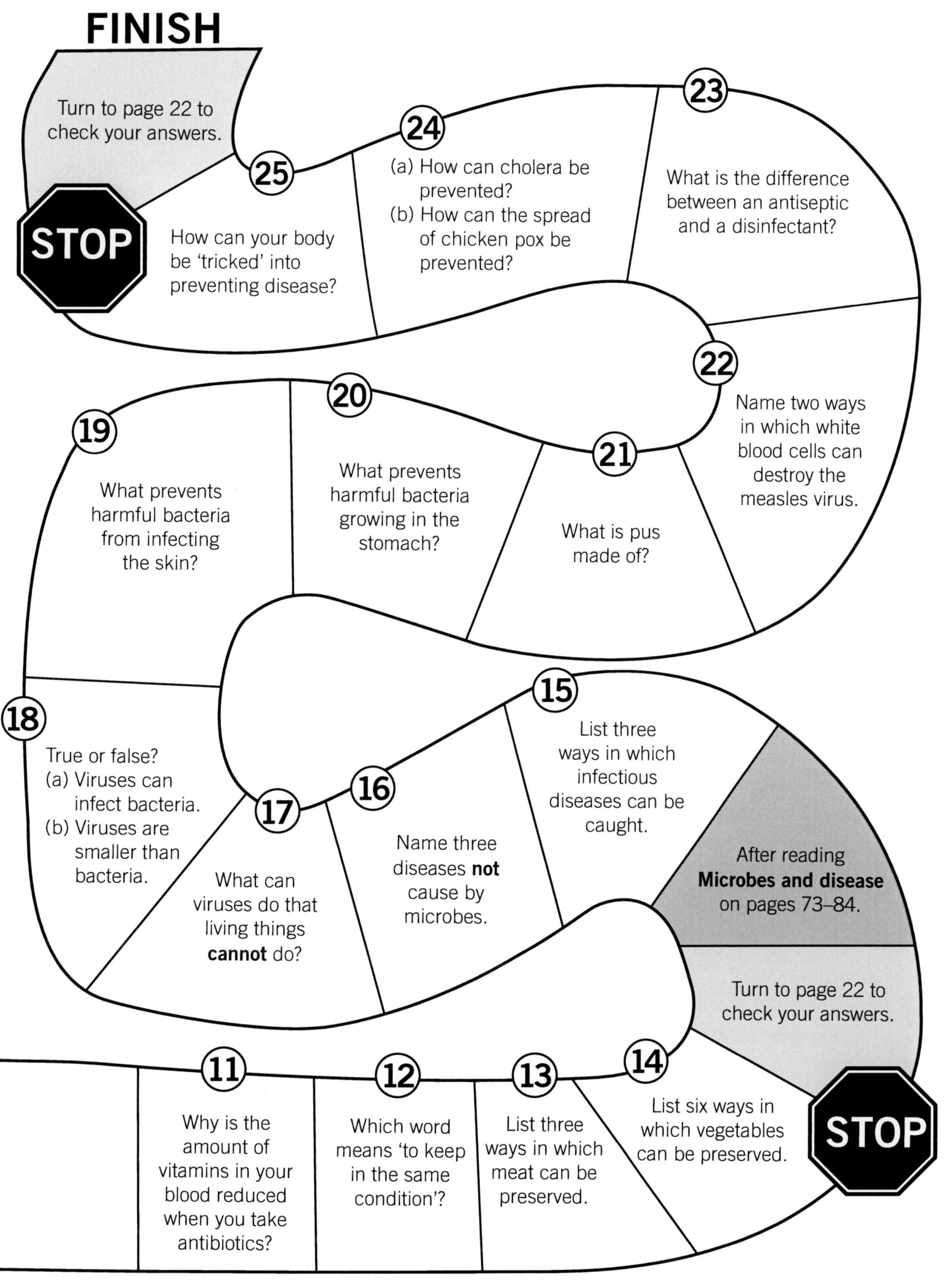
FINISH
Turn to page 22 to check your answers.
STOP
25
How can your body be 'tricked' into preventing disease?
24
(a) How can cholera be prevented?
(b) How can the spread of chicken pox be prevented?
23
What is the difference between an antiseptic and a disinfectant?
22
Name two ways in which white blood cells can destroy the measles virus.
21
What is pus made of?
20
What prevents harmful bacteria growing in the stomach?
19
What prevents harmful bacteria from infecting the skin?
18
True or false?
(a) Viruses can infect bacteria.
(b) Viruses are smaller than bacteria.
17
What can viruses do that living things **cannot** do?
16
Name three diseases **not** cause by microbes.
15
List three ways in which infectious diseases can be caught.
After reading **Microbes and disease** on pages 73–84.
Turn to page 22 to check your answers.
STOP
14
List six ways in which vegetables can be preserved.
13
List three ways in which meat can be preserved.
12
Which word means 'to keep in the same condition'?
11
Why is the amount of vitamins in your blood reduced when you take antibiotics?

Answers

Microscopic life

1 Microbes are organisms that cannot be seen without a microscope. Examples are yeast and algae.

2 (a) False
(b) True

3 A sticky coating on the cell wall.

4 (a) False
(b) True

5 Tetanus, food poisoning, cholera

Go back to page 20.

Helpful microbes

6 Water, nitrates, carbon dioxide and foul-smelling gases.

7 Yeasts

8 Fermentation

9 Mould

10 Ethanol

11 The antibiotics kill the harmless bacteria that make vitamins in the stomach.

12 Preserve

13 Refrigerating, freezing, drying

14 Heat sterilisation, refrigeration, freezing, drying, adding chemical preservatives, irradiating

Go back to page 21.

Microbes and disease

15 Physical contact with an infected person, breathing the air containing the microbes, breathing in water droplets from the air

16 Any *three* of haemophilia, asthma, multiple sclerosis, leukaemia, some heart diseases.

17 Form crystals.

18 (a) True
(b) True

19 Sweat on the skin contains salt and acids which microbes cannot tolerate.

20 Hydrochloric acid in the stomach.

21 White blood cells (phagocytes) that have engulfed bacteria and dead cells, and other liquids around the wound.

22 Antibodies and phagocytes

23 Antiseptics kill bacteria on the skin. Disinfectants kill bacteria on objects.

24 (a) hygienic conditions
(b) isolation

25 By injecting dead microbes into your blood stream, your body is tricked into making antibodies to destroy them. This then protects you when the disease really occurs.

Roundup

Exercise 5 Researching and presenting information

Choose one of the diseases listed in the table on page 74 of your textbook.

Use a library (books or internet) or your local community health centre to research one of these conditions.

Present your information on a one-page fact sheet using the format shown below (but omit the explanations in italics). Use diagrams to enhance your explanation if necessary.

Write in sentences.

Pin copies of the fact sheet around your classroom so that others can share your information.

Fact sheet on ____________________ (*state the disease*)	
Cause(s) *List the reason(s) for the disease.*	
Symptom(s) *List the change(s) in bodily function associated with this disease—the signs that this disease is occurring.*	
Effect(s) *List the result(s) of having this disease.* • short term • long term	
Treatment *Describe how the medical profession treats this disease.*	
Prevention *Describe the precautions a person could take to keep this disease from occurring.*	
Fact sheet researched and presented by ____________________ (*your name*)	

Chapter 4

Inside the atom

Overview

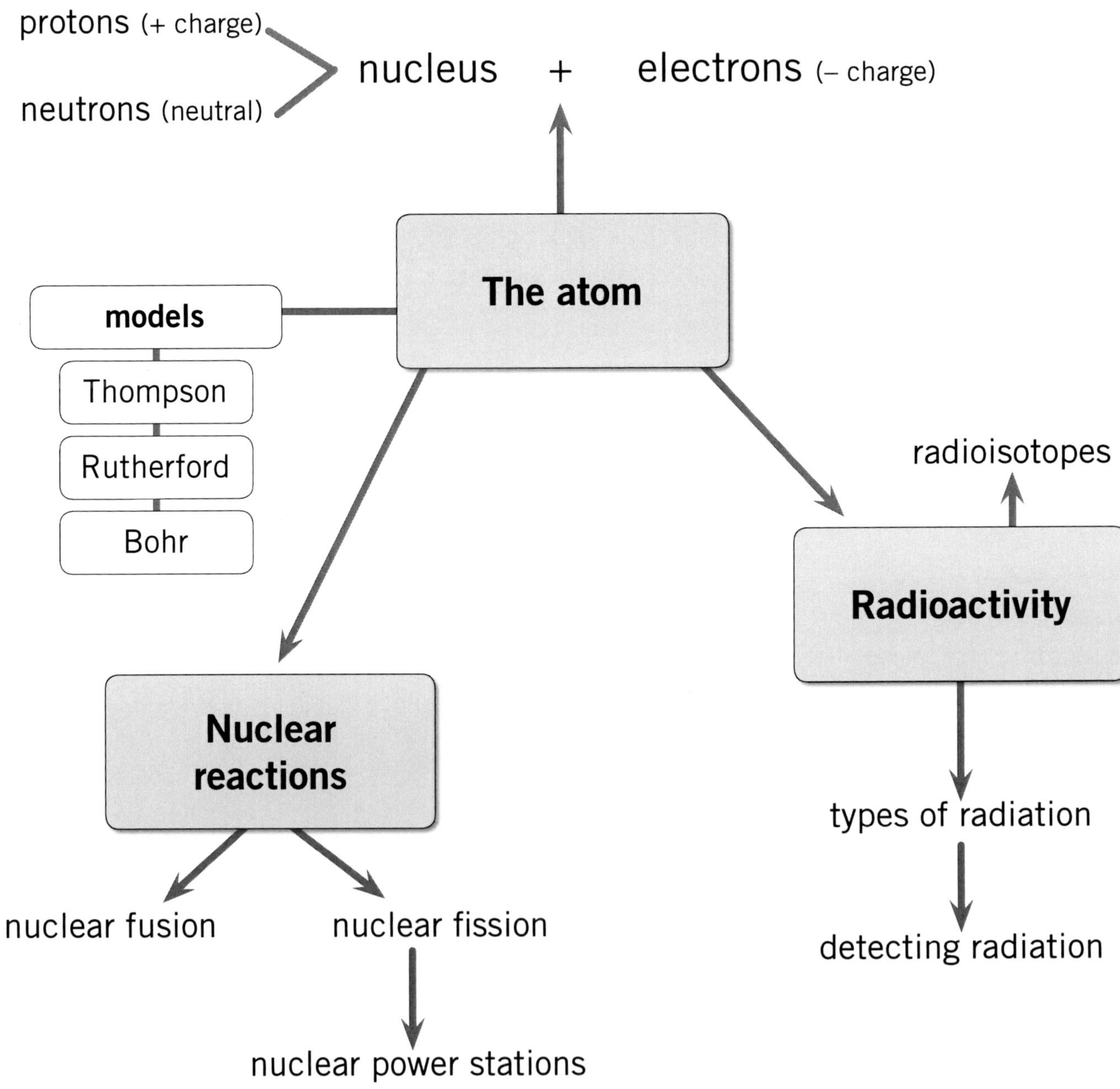

protons (+ charge)

neutrons (neutral)

nucleus + electrons (– charge)

The atom

models

Thompson

Rutherford

Bohr

Nuclear reactions

nuclear fusion

nuclear fission

nuclear power stations

Radioactivity

radioisotopes

types of radiation

detecting radiation

Exercise 1 Reflecting and sharing

Think about each of these questions individually before joining a small group to share your ideas with others.

1 Which is smallest — an atom, a proton, or a molecule?

2 Electrons are tiny, negatively-charged particles of matter that move rapidly around the nucleus of an atom. What keeps the electrons from flying away from the nucleus?

3 In a typical nuclear reaction, uranium undergoes nuclear fission. Fission means to split. What do you think occurs in a nuclear fission reaction?

4 A nuclear fission reaction releases huge amounts of energy, as in an atomic bomb. Why doesn't a nuclear power station blow up?

5 If fusion means to join, suggest how nuclear fusion is different from nuclear fission.

Keep your answers in mind as you read the chapter.

Exercise 2 Understanding linking words

Understanding the meaning of linking words is vital when reading.
Look at the difference the linking words make to the meaning of the following sentences.

Both Maria and Joel love spinach.
Unlike Maria, Joel loves spinach.

Read this paragraph carefully several times.

> Alpha, beta and gamma radiation is sometimes called ionising radiation because it can ionise atoms and molecules, that is, give them an electric charge. Generally, these ions are more likely to become involved in chemical reactions. For example, in the body these ions may cause chemical reactions that destroy cells, or they may cause uncontrolled cell growth which gives rise to tumours. However, controlled amounts of radiation can be used to treat cancer cells as these cells are more sensitive to radiation than normal cells.

1 Find a Generalisation linking word. ______________________

2 Find four Cause and effect linking words. ______________________ ______________________

______________________ ______________________

3 Find a Contrast linking word. ______________________

4 Find linking words which show that an explanation will follow. ______________________

During reading

Exercise 3 Explaining

After reading Atomic structure on pages 87–88

Sketch the faces and use your words to answer the question in each box.

Exercise 4 Making and justifying decisions

After reading Atomic structure on pages 87–89

Decide whether each statement is true or false. Circle your decision and be prepared to give reasons for (justify) your answer.

1 The nucleus of any atom contains protons and neutrons. TRUE FALSE

2 Isotopes of an element have the same chemical properties. TRUE FALSE

3 Two atoms with the same mass number of 14 must be the same element. TRUE FALSE

4 The mass number of an element is always greater than its atomic number. TRUE FALSE

5 Hydrogen-3 has three more neutrons than Hydrogen-1. TRUE FALSE

Exercise 5 Interpreting illustrations

After reading A chemical reaction on page 92

Examine the diagram on page 92 of your textbook carefully and then answer these questions.

1 Why is the red molecule larger than the grey molecule? ______________________

____

2 What is different about the molecules in the left-hand circle compared to those in the right-hand circle?

____

____

3 What is the same about the molecules in the left-hand circle compared to those in the right-hand circle?

____

____

Exercise 6 Recalling information on diagrams

After reading Nuclear power stations on page 94

Look at Fig 6 on page 94 of your textbook. Study the diagram of the nuclear power station, close your book and label the diagram with the words below.

- condenser
- steam
- steam turbine
- electric generator
- control rods
- heat exchanger
- containment building
- pump
- fuel rods

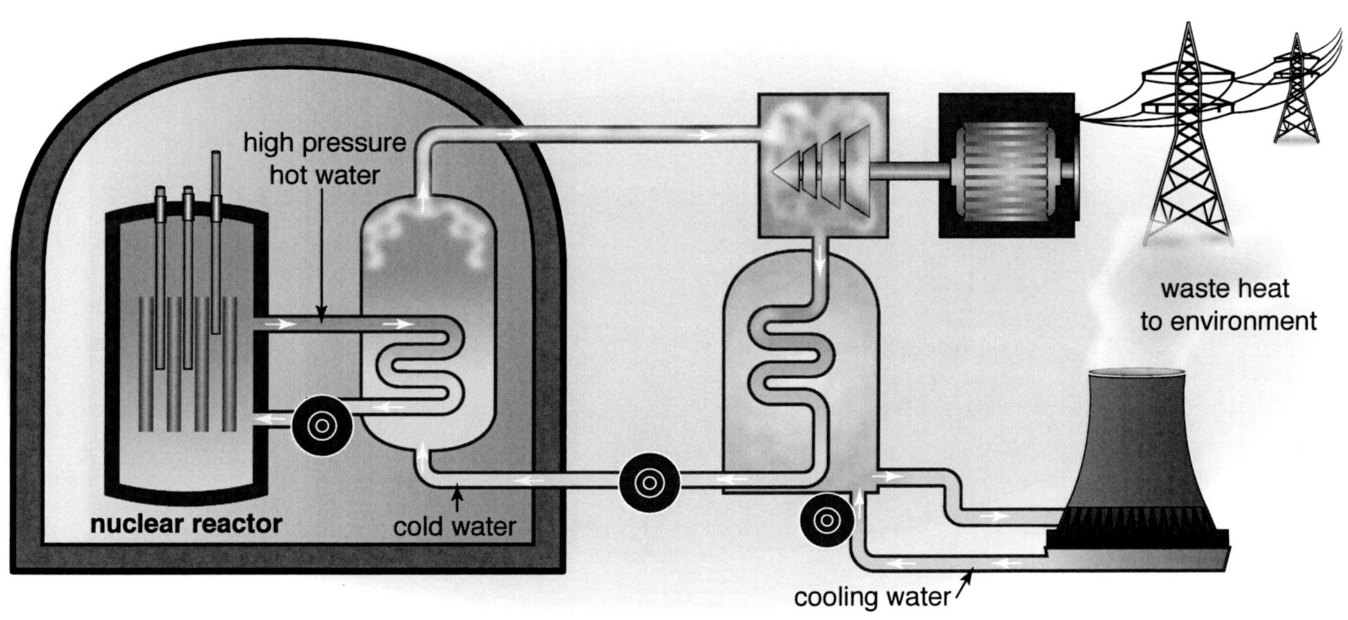

Exercise 7 Sequencing

After reading page 93

Read the following statements carefully, then place them in the correct order by placing the numerals **1 2 3 4 5** in front of them.

________ The neutron is absorbed creating a very unstable nucleus.

________ A neutron is fired at a uranium atom.

________ The reaction is called a chain reaction and releases a large amount of energy.

________ The neutrons given off from the splitting uranium are absorbed by other uranium atoms.

________ The nucleus splits into two smaller atoms and a few neutrons.

Exercise 8 Explaining

After reading Types of radiation on page 97

Complete the following table.

Type of radiation	Description	Can be stopped by ...
Alpha		
Beta		
Gamma		

Exercise 9 Recalling information

After reading Using radioisotopes on pages 99–101

1 How can cancer cells be killed using radioisotopes? ________________________________

____

2 How can radioisotopes be used to measure the thickness of paper? ________________________________

____

3 What are some other uses for radioisotopes? ________________________________

____

Exercise 10 Interpreting graphs

After reading Radioactive dating on page 102

Look at the graph on the right and answer the questions below.

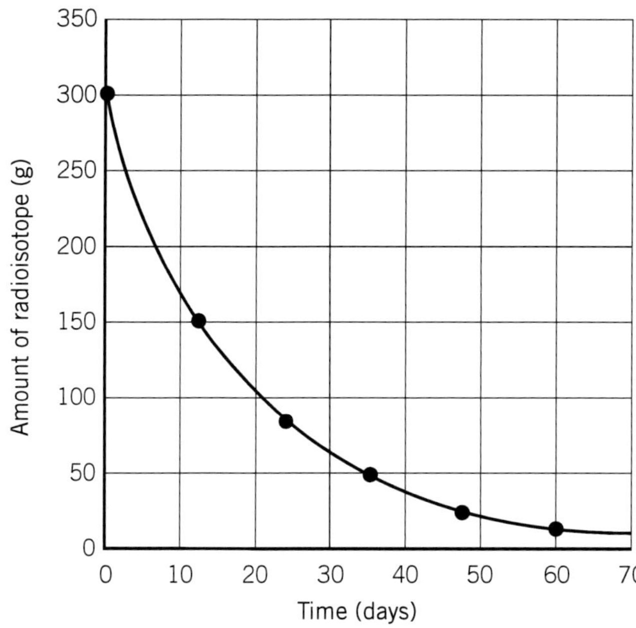

1 How many grams of the radioisotope decayed in 70 days?

____

2 What is the half-life of this radioisotope? _______________

3 How long did it take for 300 g of radioisotope to fall below 50 g? _______________

4 This radioisotope emits alpha particles. Is it suitable for detecting leaks in underground pipes? Why?

____

Roundup

Exercise 11 Reviewing your answers

Look at the questions from Exercise 1 on page 25 of your workbook.
Now that you have studied Chapter 4, write an answer for each question.

1 _______________

____

2 _______________

____

3 _______________

____

4 _______________

____

5 _______________

____

Chapter 5
Electrical energy

Overview

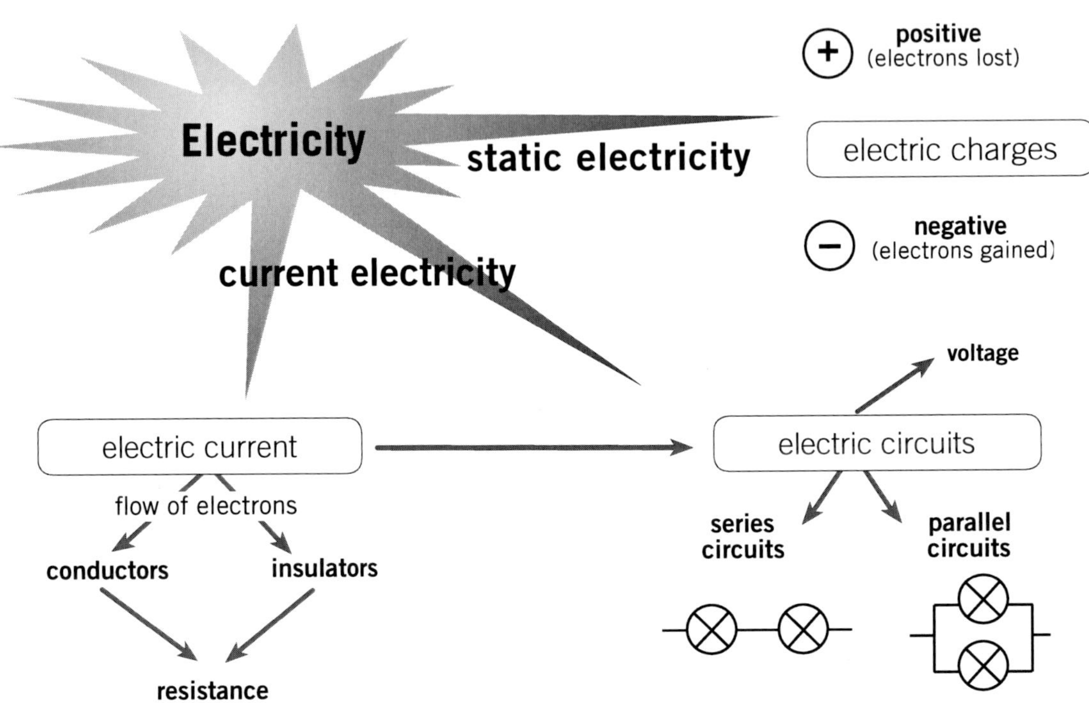

What do you know already?

Exercise 1 Exploring ideas

Form a group of three. Discuss and write responses to the following questions.

1 As you step out of the car and touch the door handle you get a small electric shock. How can you explain this?

____

____

2 Why are there two terminals on a torch battery and not just one?

____

3 What is the voltage of a torch battery? ______________________________

During reading

In this chapter you will find Core and Challenge exercises.

After reading each section of the textbook, you will find an exercise to complete. If your solution to this exercise is correct (check your answers on page 37 of this workbook), go straight to the Challenge exercise. If you cannot do the exercise, or you get it wrong, or even if you get it right but are not really sure about it, read the textbook section again and go on to the Core exercise. Of course, you may then complete the Challenge exercise.

Exercise 2

After reading Electric charges on pages 110–112

Making and justifying decisions

Some of the statements below are true and some are false. Tick the ones that are true and change the false ones to make them true.

1 There are only two types of electric charge—positive and negative.
2 Like charges attract each other and unlike charges repel each other.
3 When you rub a perspex rod with a silk cloth, electrons move from the cloth to the rod.
4 The build-up of electric charges on objects is called current electricity.
5 Two ebonite rods rubbed with wool repel each other.
6 In a thunderstorm, the tops of the clouds are normally positive and the bottoms negative.

Check your answers on page 37 of this workbook.

Core exercise

Inferring

You have four charged rods—A, B, C and D suspended as shown.

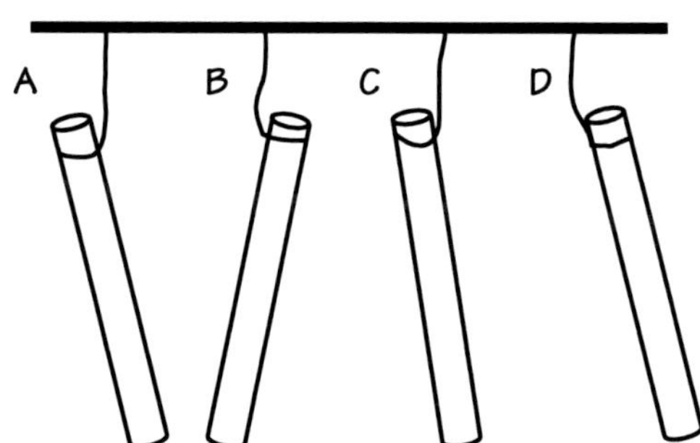

If you know B has a positive charge, what charge does each of the others have?

A ______________________

C ______________________

D ______________________

Challenge exercise

Explaining ideas

Complete the cartoon to help Kate understand this section.

I understand about positive and negative charges but what's that got to do with atoms?

It's like this Kate, ______________________

____

____

____

____

Exercise 3

After doing Investigation 10 on page 117

Labelling

On the diagram, label the following items:

- alligator clip
- torch cell
- cell holder
- bulb
- bulb holder
- connecting wire
- switch

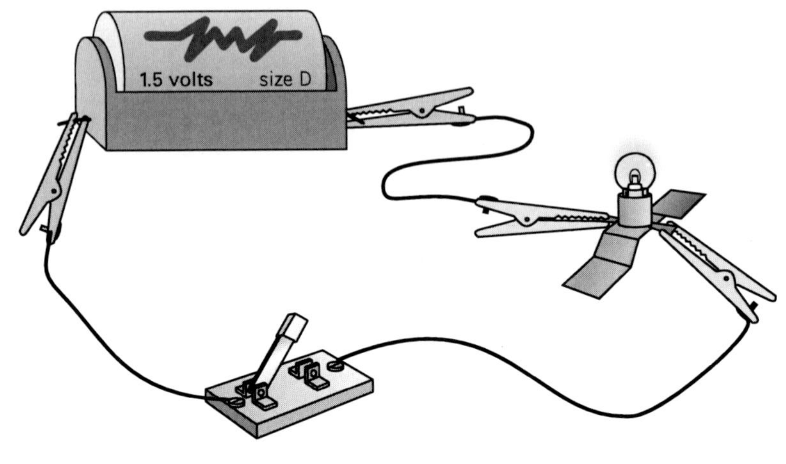

Check your answers on page 37 of this workbook.

Core exercise

Making and justifying decisions

Zlatko says you need only one wire to make a torch bulb glow. Heidi says you need two wires.

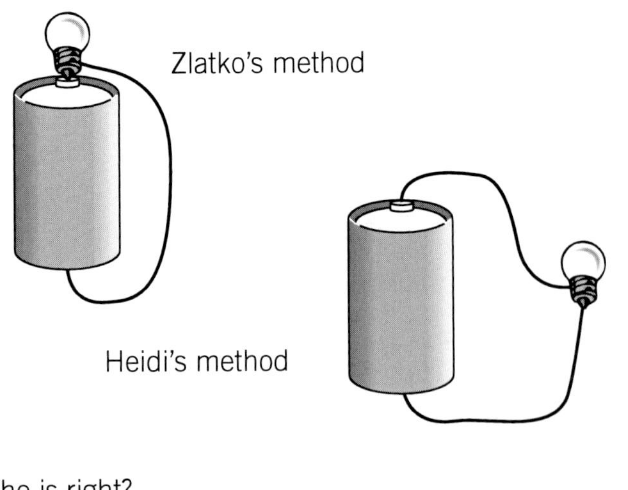

Who is right? ______________________

Why? ______________________

____

____

____

Challenge exercise

Looking for alternatives

In the Core exercise, Zlatko says he knows another way to make the bulb glow with just one wire. Sketch how this could be done.

Exercise 4

After reading What is a circuit? on page 118

Using analogies

Complete the table on the right which compares an electric circuit to water flowing through pipes. The first two have been done for you.

water flowing in pipes	electric circuit
pump	cell
pipes	
fountain	
water	
water meter	

Check your answers on page 37 of this workbook.

Core exercise

Completing sentences

Use the diagrams on page 118 to help you fill in the missing words in the following sentences.

If you put in a bigger fountain, the rate of flow will decrease. If you want to increase the flow rate you need a bigger

____________________.

Similarly, if you want a bulb to glow more brightly you need a

____

with more push. This 'electrical push' is called its

____________________,

and is measured in

____________________.

Challenge exercise

Exploring ideas

A group of students made a torch bulb glow by connecting it to a cell with two wires. They then discussed how the electric current flows through the wires.

1 Who do you think is right? ____________________

Why?____________________

____

____

2 Suggest a way of checking who is right. ____________________

____

____

Exercise 5

After doing Investigation 11 on page 119 and reading pages 120–121

Interpreting data

A group of students did Investigation 11 and obtained these results.

Object/material	Does bulb glow?	Ammeter reading (milliamps)
steel nail	✓	100
piece of string	✗	5
plastic rod	✗	–
aluminium foil	✓	150
copper coin	✓	200

1 Which object is the best conductor?

____

2 Which object is the best insulator?

____

3 Which of the five materials would be best for covering the handle of a pair of electrician's pliers?

____

4 Which material would be best for making electrical power lines? ______________________

Check your answers on page 37 of this workbook.

Core exercise

Defining and generalising

1 Complete these definitions.

a A conductor is a material through which

____

____

An example is ______________________

b An insulator is a material through which

____

____

An example is ______________________

2 Cross out one word in each of the brackets to make these generalisations true.

a The (better/worse) the conductor, the (higher/lower) the resistance.

b In a (conductor/insulator) the electrons (can move easily/cannot move).

Challenge exercise

Designing an experiment

Design a simple experiment to find out which of the following metals has the lowest resistance.

In your design, make sure you mention which variables you will need to control.

steel　　tin　　aluminium　　copper

Exercise 6

After reading Electric circuits on pages 124–125

Interpreting diagrams

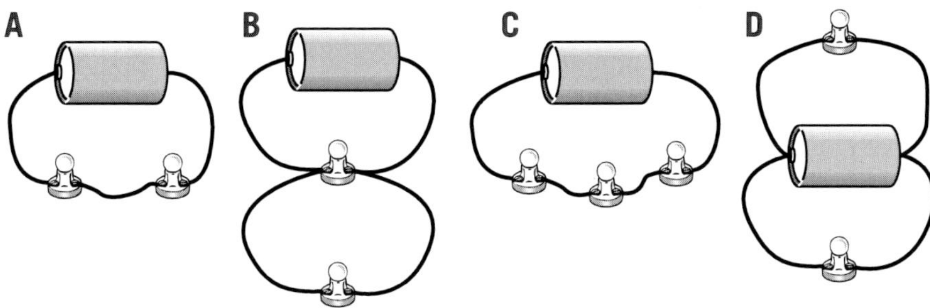

1 In which circuits are the bulbs connected in series? ____________

2 In which circuits are the bulbs connected in parallel? ____________

3 In which circuits would the bulbs be brightest? ____________

4 In which circuit would the bulbs be dimmest? ____________

Check your answers on page 41 of this workbook.

Core exercise

Interpreting diagrams

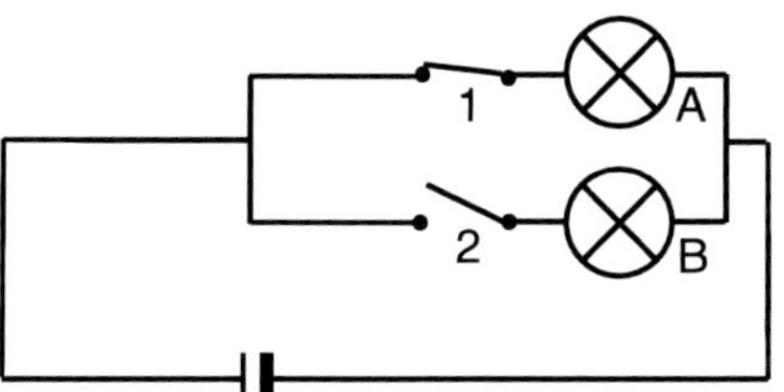

1 Here is a circuit diagram.

Which bulb is glowing? ____________

What will happen if I close Switch 2 as well?

____

____

2 Here is a second circuit.

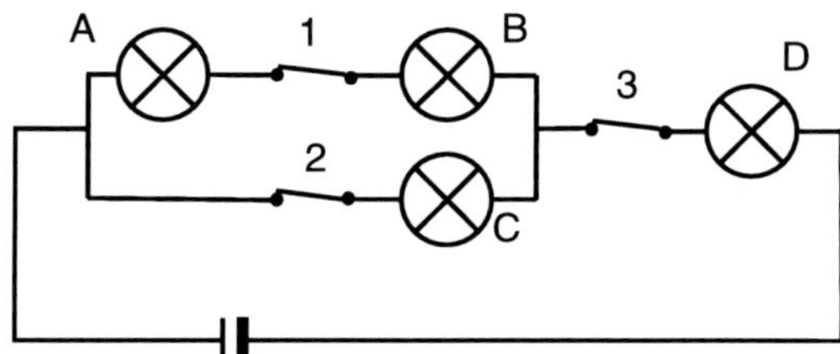

If I open Switch 1, which bulbs go out? ____________

If I open Switch 2, which bulbs go out? ____________

If I open Switch 3, which bulbs go out? ____________

Challenge exercise

Solving a problem

Design a circuit diagram for the following situation.

A caravan has 4 lights—one over the kitchen sink, one in the living area, one in the bedroom and one outside light. Each light operates independently, except the outside light, which comes on when the living area light is switched on.

Make sure your diagram includes a battery and 3 switches.

Exercise 7 Explaining

After doing Investigation 12 on pages 125–127

Interpreting diagrams

For each circuit below cross out *will* or *will not*, then say why.

A

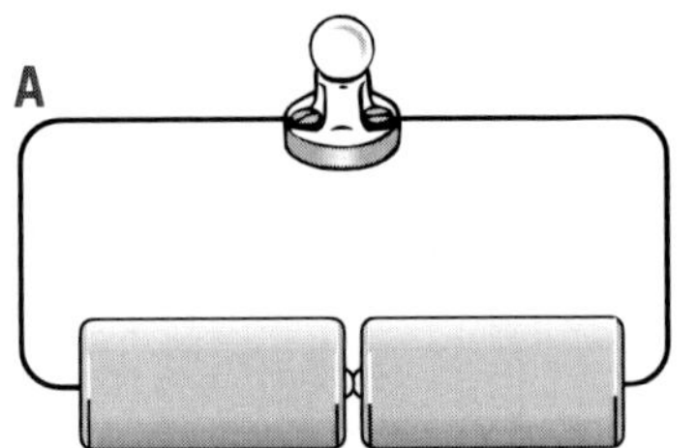

The bulb will/will not

glow because ______________

____

____

B

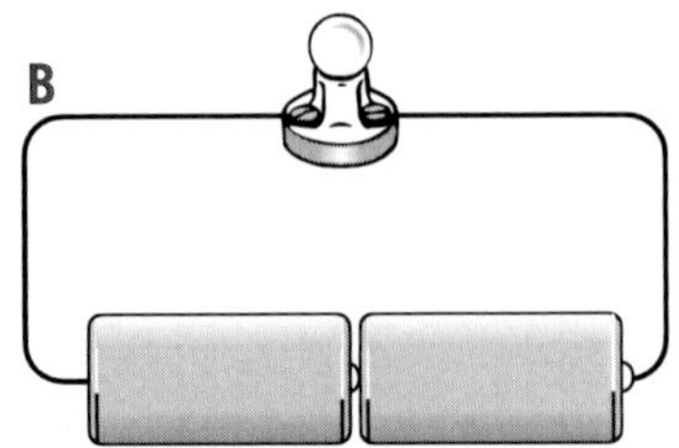

The bulb will/will not

glow because ______________

____

____

C

The bulb will/will not

glow because ______________

____

____

Check your answers on page 37 of this workbook.

Core exercise

Interpreting diagrams

1 Blake has put the batteries into his stereo as shown, but it doesn't work. Circle the battery he has put in wrongly.

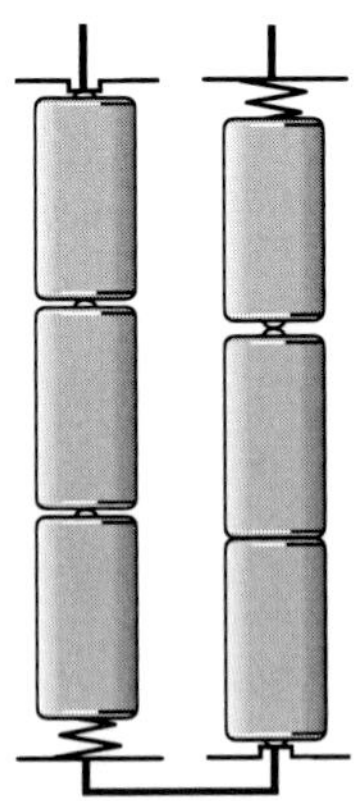

2 Are the batteries connected in series or in parallel?

____

3 If each of the batteries is 1.5 volts, what is the total operating voltage of the stereo?

______________ volts

Challenge exercise

Interpreting diagrams

1 Carrie sets up the circuit shown, but one of the connecting wires breaks.

Will the bulb still glow? ______________

Explain your answer ______________

____

____

2 The bulbs in Circuit B glow more dimly than the bulbs in Circuit A.

A

B

How could you make the bulbs in Circuit B glow as brightly as the bulb in Circuit A? Draw a diagram below.

Answers

Exercise 2

1 TRUE
2 FALSE Like charges *repel* and unlike charges *attract*.
3 FALSE When you rub a perspex rod with a silk cloth, electrons move from the *rod* to the *cloth*.
4 FALSE The build-up of electric charges on objects is called *static electricity*.
5 TRUE
6 TRUE

Exercise 3

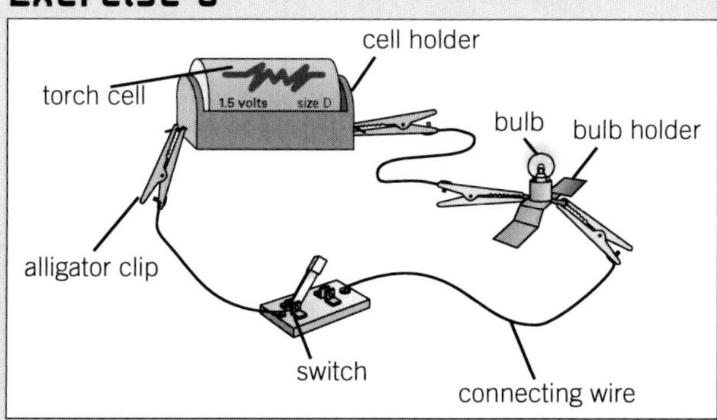

Exercise 4

water flowing in pipes	electric circuit
pump	cell
pipes	connecting wires
fountain	bulb
water	electric current or electrons
water meter	ammeter

Exercise 5

1 copper coin
2 plastic rod
3 plastic
4 copper

Exercise 6

1 A and C
2 B and D
3 B and D
4 C

Exercise 7

A The bulb *will not* glow because the two positive terminals of the cells are connected.
B The bulb *will* glow because the positive terminal of one cell is connected to the negative terminal of the other.
C The bulb *will not* glow because the two negative terminals of the cells are connected.

Roundup

Exercise 8 Clarifying ideas and concepts

Merryl has found this diagram of how a torch works, but she doesn't understand it. Help her by completing the conversation below.

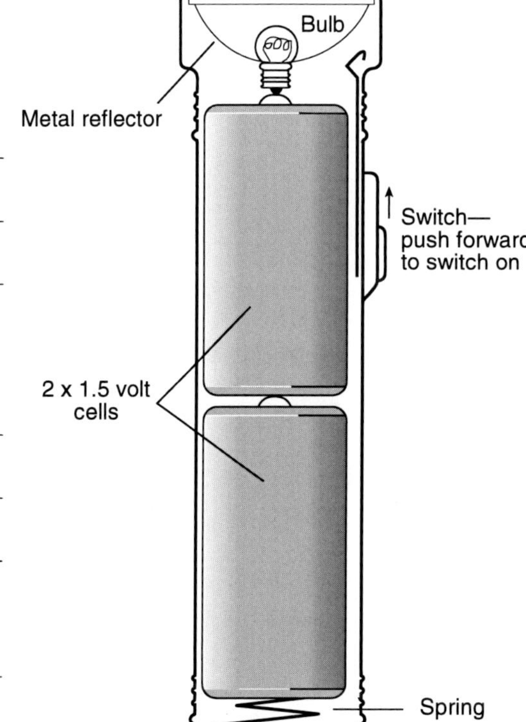

Merryl: I can't see how pushing the switch forward makes anything happen.

You:____

Merryl: I suppose you're right, but there's no wire from the bottom cell to the bulb.

You:____

Merryl: I see—and I suppose the purpose of the spring is …

Chapter 6

Everyday reactions

Overview

Acids

- release hydrogen ions (H^+) in solution
- have a pH less than 7 on the pH scale
- acid + metal ⟶ salt + hydrogen gas
- acid + carbonate ⟶ salt + water + carbon dioxide gas
- carbon dioxide + water ⟶ carbonic acid (acid rain)

Bases

- release hydroxide ions (OH^-) in solution
- have a pH more than 7 on the pH scale

Both Acids and bases

- are important in everyday life
- neutralise each other (acid + base ⟶ salt + water)

What do you know already?

Exercise 1 Explaining

Do these before you read the chapter.

Why do you burp after drinking a fizzy drink?

Why do teeth decay?

During reading

Working independently

In this chapter of the workbook you will work independently at your own pace.
You will be playing Acid and Base-ball. The rules are simple.

RULES

1. You start at Home Base and read the section of your textbook indicated on the inside of the baseball diamond.
2. You complete all the exercises between each base and the next. When you finish each exercise, shade in that square to show your progress around the baseball diamond. (There are four exercises between each base and the next.)
3. When you reach each base, turn to page 47 to correct your work.
   Put your tally on the scoreboard in the centre of the diamond.
4. If you score 4 out of 4, proceed to the next base. If not, go back and work out why you were wrong (your teacher can help you here) and repeat the exercise.
   You must not move to the next base until you have all the exercises correct.
5. Continue at your own pace around the diamond until you reach Home Base.

ACID AND BASE-BALL DIAMOND

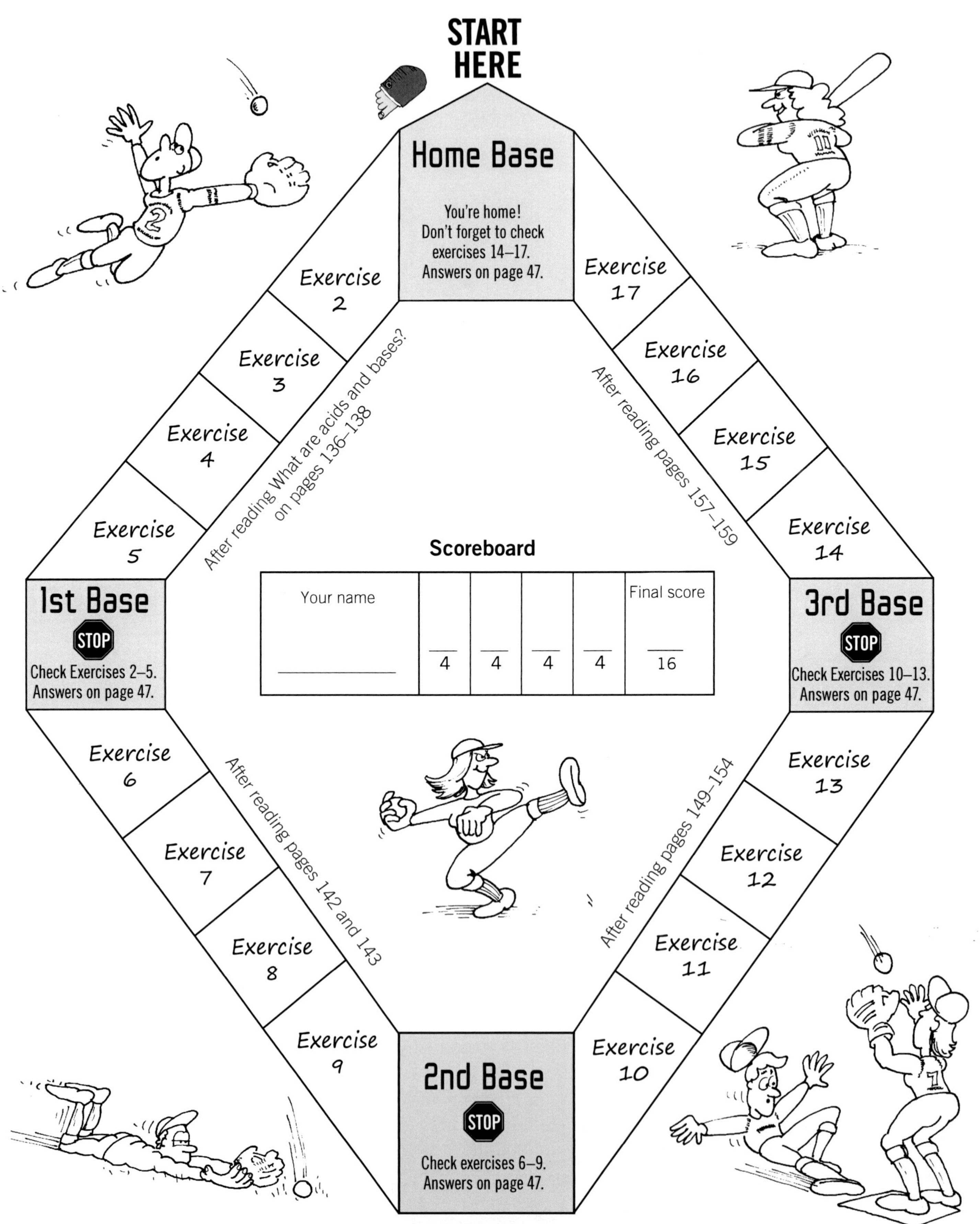

Your name					Final score
______	/4	/4	/4	/4	/16

Home Base to First Base

After reading What are acids and bases on pages 136–138

Exercise 2 Comparing and contrasting

After studying the photo on page 136

1 Complete the sentence comparing cream of tartar and grapes.

Both ____.

2 Complete the sentence contrasting yoghurt and vinegar.

______________________ but ______________________.

3 Complete the sentence comparing and contrasting oranges and rhubarb.

Both ____, however

______________________ while ______________________.

Exercise 3 Completing word equations

Complete these word equations.

1 food particles + bacteria + [ ] + [ ] = plaque

2 fine abrasive + mild detergent + fluoride compounds = [ ]

3 [ ] + [ ] + pepsin + water = gastric juice

4 lack of mucus + stomach acid = [ ]

5 [ ] + cream of tartar + moisture = carbon dioxide gas

Exercise 4 Matching in a magic square

Match each word with its function.

Put the number of the function in the appropriate lettered magic square.
If you have the answers right, the sum of the numbers across, down and diagonally will be the same.

	WORD		FUNCTION
A	toothpaste abrasive	**1**	protects the outside of your teeth
B	enamel	**2**	causes bread to rise when baked
C	acetic acid	**3**	makes tooth enamel more resistant to acids
D	fluoride	**4**	breaks down proteins into amino acids
E	gastric juice	**5**	breaks down food in the stomach
F	hydrochloric acid	**6**	preserves food in pickles and sauces
G	enzymes	**7**	kills microbes in the stomach
H	mucus	**8**	scrapes food particles from teeth
I	carbon dioxide gas	**9**	protects the stomach wall from acid

A	**B**	**C**
D	**E**	**F**
G	**H**	**I**

Exercise 5 Applying a series of steps to reach a conclusion

A student has been working with an acidic solution, a basic solution and two of the indicators listed on page 138 of your textbook. They are all colourless except for one which is red. They are in separate bottles A, B, C and D, but the student cannot remember which is the acid, which is the base, and which are the indicators.

To try to work out which solution is which, she tested small samples of the solutions in test tubes. Here are her observations:

Solution A + Solution B	Solution A + Solution C	Solution B + Solution C	Solution A + Solution D	Solution B + Solution D	Solution C + Solution D
colourless	colourless	pink	red	red	blue

Use the results to work out which solution is acidic, which is basic and the names of the two indicators.

Working space

Solution A is ______________________ Solution C is ______________________

Solution B is ______________________ Solution D is ______________________

You have reached first base.
Turn to page 47 to check your answers.

First Base to Second Base

After reading The pH scale on pages 142–143

Exercise 6 Making a reasonable guess

Think about the composition of acids and bases then make a reasonable guess.
Start by eliminating the least likely possibilities. The letters pH stand for—

1 phenolphthalein **2** paper hypothesis **3** power of hydrogen **4** phenol hydrogen

My reasonable guess is ______________________

Exercise 7 Making and justifying decisions

Read the statements below and decide if each is true or false.
Circle your answer and be able to justify this decision.

1	The higher the pH, the more acidic the solution.	TRUE	FALSE
2	Neutral solutions have a pH of 0.	TRUE	FALSE
3	A solution which changes pH paper to a red colour is acidic.	TRUE	FALSE
4	Some toothpaste is moderately basic so that it can neutralise the effect of acidic saliva on teeth.	TRUE	FALSE
5	Lemons have a pH of 10.	TRUE	FALSE

Exercise 8 Writing cause and effect sentences

Use Cause and effect linking words from page 92 to make one sentence from each pair of sentences.

1 a The pH of your stomach is about pH 2.
  b The digestion of proteins in the stomach requires acidic conditions.

____

____

2 a The soil in the paddock is too alkaline.
  b Adding compost or manure will lower the pH of the soil.

____

____

Exercise 9 Applying information

Mr and Mrs Jones have a farm with nine fields. The pH of the soil in each field is shown in the diagram.

Field 1 pH 5.0	Field 2 pH 7.5	Field 3 pH 5.0
Field 4 pH 4.5	Field 5 pH 9.0	Field 6 pH 7.0
Field 7 pH 5.0	Field 8 pH 8.5	Field 9 pH 5.0

1 Use coloured pencils to shade the fields that contain acidic soils in red, and those that contain basic soils in blue.

2 In Field 5, Mr Jones wants to plant roses, which require a soil with a pH of 6.0–7.0. How can he make the soil more suitable?

____

____

3 Mrs Jones wants to plant beetroot, which requires soil with a pH between 7.0 and 8.0. In which fields can she plant the beetroot? What would she have to do to the soil to plant the beetroot in Field 4?

____

____

____

You have reached second base.
Turn to page 47 to check your answers.

Second Base to Third Base

After reading Reaction of acids and bases on pages 149–154

Exercise 10 Completing word equations

Complete these word equations.

1 copper + sulfuric acid ⟶ copper sulfate + ______

2 calcium carbonate + hydrochloric acid ⟶ calcium chloride + ______ + ______

3 ______ + ______ ⟶ potassium nitrate + hydrogen

4 sodium + ______ ⟶ ______ + hydrogen

5 ______ + ______ ⟶ magnesium chloride + hydrogen

Exercise 11 Explaining in simple terms

Complete the cartoon below using your own words. Don't copy from the textbook.

Exercise 12 Using the correct word

Insert the correct form of the word 'neutral' from the box into the following sentences.

neutral
neutralise
neutralisation
neutraliser
neutralised

1 Baking soda can be ______ by adding an acid.

2 When an acid and base react to produce a solution that is neither acidic nor basic, the reaction is known as ______ .

3 Calamine lotion will ______ the effects of a bee sting.

4 Water is a ______ solution.

5 After a tint, which is alkaline, a hairdresser applies a ______ to your hair.

Exercise 13 Devising solutions

List three practical things that Australians could do to reduce the problem of acid rain in our environment.

1 ______________________________

____

____

2 ______________________________

____

____

3 ______________________________

____

____

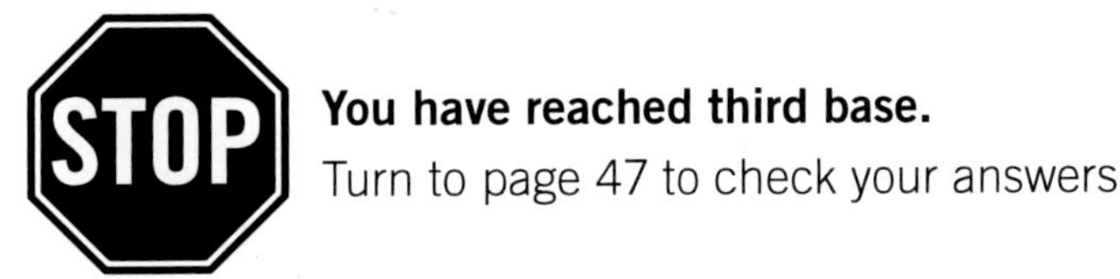

You have reached third base.

Turn to page 47 to check your answers.

Third Base to Home Base

After reading Energy in reactions pages 157–159

Exercise 14 Recalling information

Insert the missing words from those in the box.

exothermic	glucose	digesting	carbon	dioxide	endothermic	energy

1 The simple sugar ____________ is produced in the process of photosynthesis.

2 Reactions that need energy to make them go, like photosynthesis are called ______________.

3 During cellular respiration energy is released and ______________ is also produced.

4 The combustion of petrol in an engine requires ______________ usually in the form of a spark to start the reaction.

5 Animals use some of the energy produced in cellular respiration in ______________ foods.

Exercise 15 Inferring

The text says 'Life on Earth would not be possible without photosynthesis. It is vital for producing oxygen in the atmosphere'.

1 Where does all the oxygen in the atmosphere come from?

____

____

2 Why is photosynthesis described as an endothermic reaction?

____

____

Exercise 16 Comparing and contrasting

Read pages 158 and 159 carefully. Use the Compare and contrast linking words on page 92 to complete sentences for the two questions below. Describe both similarities and differences for each.

1 Photosynthesis and respiration are *different* in that ______________________

____

____

2 Respiration and combustion ______________________

____

____

Exercise 17 Writing in your own words

1 In your own words, describe what ozone is and how it protects living things on Earth.

____

____

2 In your own words, describe how CFCs can increase the amount of UV radiation reaching the Earth.

____

____

You have reached home base.
Turn to page 47 to check your answers.

Answers

Home Base to First Base

Exercise 2

1 Both cream of tartar and grapes contain tartaric acid.
2 Yoghurt contains lactic acid but vinegar contains acetic acid.
3 Both oranges and rhubarb contain acids, *however* oranges contain citric acid *while* rhubarb contains oxalic acid.

Exercise 3

1 acid + saliva
2 toothpaste
3 hydrochloric acid + rennin
4 stomach ulcer
5 baking soda (or sodium hydrogen carbonate).

Exercise 4

A 8	B 1	C 6
D 3	E 5	F 7
G 4	H 9	I 2

Exercise 5

The colourless—pink indicator is phenolphthalein (probably B).
A + B = colourless ⟶ A is acidic
B + C = pink ⟶ C is probably basic
A + C = colourless since no indicator
Red—blue indicator is probably litmus (probably D)
A + D = red ⟶ confirming A is acidic
B + D = red since D is red
C + D = blue ⟶ confirming C is basic
So: Solution A is acidic
Solution B is phenolphthalein indicator
Solution C is basic
Solution D is litmus indicator

Put your score on the scoreboard and continue with Exercises 6–9 after correcting any answers that were wrong.

First Base to Second Base

Exercise 6

3 power of hydrogen

Exercise 7

1 FALSE 2 FALSE 3 TRUE 4 TRUE 5 FALSE

Exercise 8 *(Your answer should be something like this)*

1 The pH of your stomach is about pH 2 because the digestion of proteins requires acidic conditions.
2 If the soil in the paddock is too alkaline, then adding compost or manure will lower the pH of the soil.

Exercise 9

1 The acidic fields are 1, 3, 4, 7 and 9.
The basic fields are 2, 5 and 8. (Field 6 is neutral.)
2 Add compost, manure or a soluble fertiliser such as ammonium nitrate.
3 Fields 2 and 6 are suitable for beetroot.
Powdered limestone or dolomite could be used on Field 4.

Put your score on the scoreboard and continue with Exercises 10–13 after correcting any answers that were wrong.

Second Base to Third Base

Exercise 10

1 neutral 2 ion 3 lose, positive
4 non-metals, negative 5 ionic bonds

Exercise 11

1 … then H^+Cl^- dissolved in water is acidic.
2 … then Na^+OH^- dissolved in water is basic (or alkaline).
3 … then H^+ ions cause the sharp biting feeling on your skin.
4 … then distilled water does not contain ions.

Exercise 12

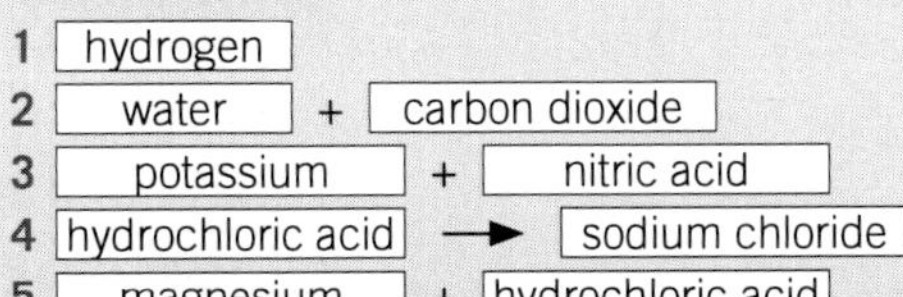

1 hydrogen
2 water + carbon dioxide
3 potassium + nitric acid
4 hydrochloric acid ⟶ sodium chloride
5 magnesium + hydrochloric acid

Exercise 13 *(Your answer should be something like this)*

Put your score on the scoreboard and continue with Exercises 14–17 after correcting any answers that were wrong.

Third Base to Home Base

Exercise 14

1 glucose 2 endothermic 3 carbon dioxide
4 energy 5 digesting

Exercise 15

1 It is produced in the process of photosynthesis carried out by green plants and phytoplankton.
2 The process requires energy to combine water and carbon dioxide, and to produce glucose and oxygen.

Exercise 16

1 Photosynthesis and respiration are *different* in that photosynthesis requires energy *whereas* respiration releases energy. *However, both* processes occur in living things.
2 Respiration and combustion both release energy. *However,* respiration is a controlled release *while* respiration is an uncontrolled releases.

Exercise 17

1 Ozone forms when an atom and a molecule of oxygen combine. Ozone absorbs UV radiation and stops it from reaching the Earth's surface.
2 The UV radiation breaks apart CFCs releasing chlorine which, in turn, reacts with ozone, breaking it up. With less and less ozone, the UV radiation can get through to the Earth's surface.

Put your score on the scoreboard and tally your final score. Don't forget to correct any answers that were wrong.

Chapter 7
Body balance

Overview

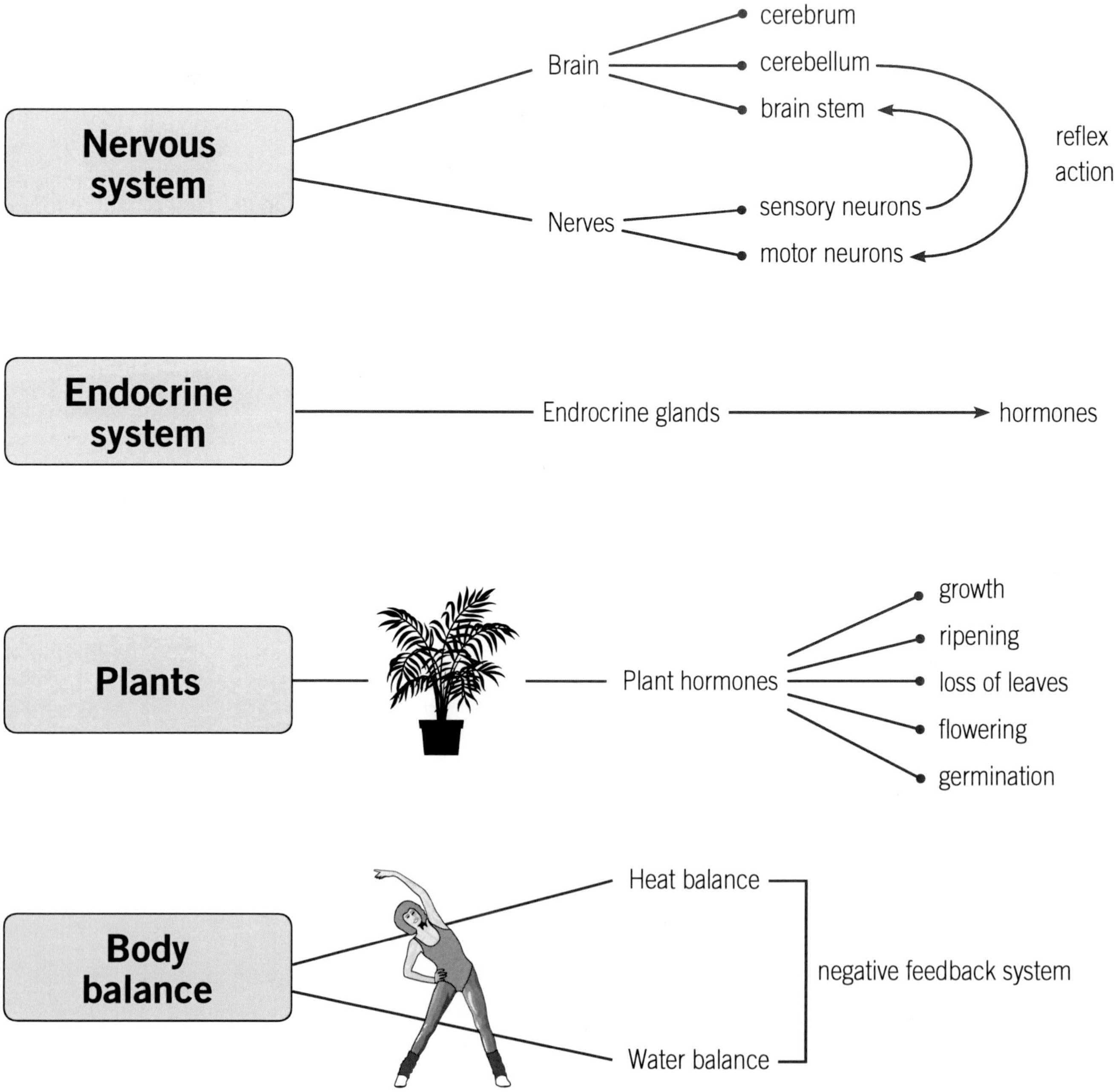

Nervous system
- Brain
  - cerebrum
  - cerebellum
  - brain stem
- Nerves
  - sensory neurons
  - motor neurons

reflex action

Endocrine system
- Endrocrine glands → hormones

Plants
- Plant hormones
  - growth
  - ripening
  - loss of leaves
  - flowering
  - germination

Body balance
- Heat balance
- Water balance

negative feedback system

What do you know already?

Exercise 1 Checking prior knowledge

Link the body part with its function.

1 brain stem	• produces insulin
2 pituitary gland	• filter wastes from the blood
3 sweat glands	• responsible for involuntary actions e.g. heartbeat, pulse, digestion, breathing
4 cerebrum	• reduce temperature
5 pancreas	• controls memory, speech, walking
6 kidneys	• produces growth hormone

When you have completed this chapter, come back to this exercise to check your answers.

Exercise 2 Understanding words in context

As well as understanding scientific words, you also need to understand general vocabulary so that you can make sense of information.

You can often work out the meaning of an unknown word by reading the words around it (the context). Below are six words in italics that are used in this chapter. Use the context clues to try to work out their meanings.

The words in context	Your guess at the meaning (in your own words)
Even though she was supposed to be silent, she let out an *involuntary* gasp when she saw the spider.	
The smell of bacon frying in the pan always *stimulates* her appetite.	
When he had finished filleting the fish, he put the knife back in the *sheath* attached to his belt.	
He was sick of everyone treating him like a little kid. He couldn't wait to reach *puberty.*	
It didn't take long for the sponge to *absorb* the water on the bench.	
When the detectives had *eliminated* all other suspects, they arrested her for murder.	

During reading

Exercise 3 Recalling information

After reading Nerves and hormones on pages 165–166

Complete the table below by reading the statement and deciding to which part(s) of the brain this statement applies. If it applies, place a ✔; if it does not apply, place a ✗. The first one is done for you as an example.

Note: Some statements may apply to more than one part of the brain.

Statement	Part of brain Cerebrum	Cerebellum	Brain stem
1 This is the largest part of the brain.	✔	✘	✔
2 This part controls thought and reasoning.	______	______	______
3 This part controls balance and posture.	______	______	______
4 This part controls voluntary actions.	______	______	______
5 This part controls involuntary actions.	______	______	______
6 This part controls breathing, heartbeat, pulse and digestion.	______	______	______
7 This part controls memory.	______	______	______
8 This part receives information from your senses.	______	______	______
9 This part helps you balance and co-ordinates muscles.	______	______	______
10 This part controls actions such as walking, running and jumping.	______	______	______

Exercise 4 Sequencing

After reading pages Types of nerves on page 167 and Reflex action on 169

Rearrange the events in the correct sequence by placing the numbers 1–5 in the boxes.

- Your brain sends a message via motor neurons to your muscles. ☐
- A nerve impulse is sent via sensory neurons to the spinal cord and brain. ☐
- Your muscles move your legs for a quick getaway. ☐
- Your eyes see a bull charging towards you. ☐
- Your spinal cord and brain receive the message. ☐

Exercise 5 Researching and presenting a fact sheet

After reading Nerves and hormones on pages 165–172

Task: In pairs, research and present a fact sheet on any one of the following disorders—

Nervous system disorders: neuralgia, neuritis, paraplegia, sciatica, quadriplegia

Endocrine system disorders: acromegaly, Addison's disease, Cushing's disease, goitre, myxoedema, osteoporosis, thyrotoxicosis

Your fact sheet should be a single sheet of A4 paper with information in the same format as shown on page 23 of this workbook. Use the table below for writing notes. You should use more than one source to gather information. Where appropriate, use simple diagrams to illustrate what you write.

1 Cause(s)	
2 Symptoms(s)	
3 Effects(s) • short term • long term	
4 Treatment	
5 Prevention	

Pin copies of your fact sheet around the room for others to share.

Exercise 6 Completing a cause and effect table

After reading Responses in plants on pages 175–177

Throughout the chapter, there are many examples of cause and effect.

In your textbook are several examples of how plants react to stimuli. Complete the table.

Cause	Effect
1	Some plants close their leaves at night and open them during the day.
2	When a seed germinates, the root grows downwards (even if it is turned upside down).
3 Plants contain a group of hormones called auxins.	
4	Ripe fruits taste sweeter than unripe fruits.

Exercise 7 Writing cause and effect sentences

After completing Exercise 6 above

Choose two of the pairs of causes and effects and join each pair into one sentence.
Use the Cause and effect linking words from page 92.

1 ____

____

2 ____

____

Exercise 8 Applying information

After reading Heat balance on pages 179–180

Suppose you and two friends are bushwalking in the Snowy Mountains when you become lost. The temperature drops to 3°C. Suggest five ways you could prevent heat loss from your body.

1 ______________________________

____

2 ______________________________

____

3 ______________________________

____

4 ______________________________

____

5 ______________________________

____

Exercise 9 Writing in role

After reading Water balance on pages 181–183

Write a short response to the following question in a magazine. (You are the doctor.)

> Dear Doctor,
>
> All the health magazines say that I should drink at least eight glasses of water per day. I hate water! Why do I need to have so much?
>
> Waterlogged Wendy

Dear Waterlogged Wendy,

____

____

____

____

____

____

____

____

Roundup

Exercise 10 Reviewing your prior knowledge

Go back to Exercise 1 to see if you can now improve the meanings of the words.

Exercise 11 Writing a cause and effect essay

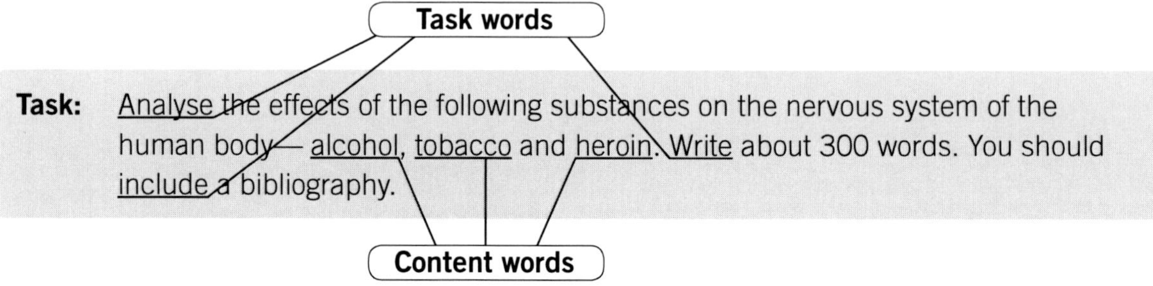

1 Understanding the task
Read the task several times, thinking about the task and content words. Then check with a partner that you understand exactly what to do. (Make sure you know what makes up the human nervous system.)

2 Preparing for research
The best way to gather relevant information is to draw up a table using the content words of the task. (Your table will need to be much larger than the one below.)

Cause	Effect (What does this substance do to the body's nervous system?)
Alcohol	
Tobacco	
Heroin	

3 Finding the relevant information
Use the internet or your library to find information on the three substances.

As you research using different sources, list them as shown in the diagram.

When you have completed your assignment, add this list of sources (in alphabetical order, according to author) to the assignment as a bibliography.

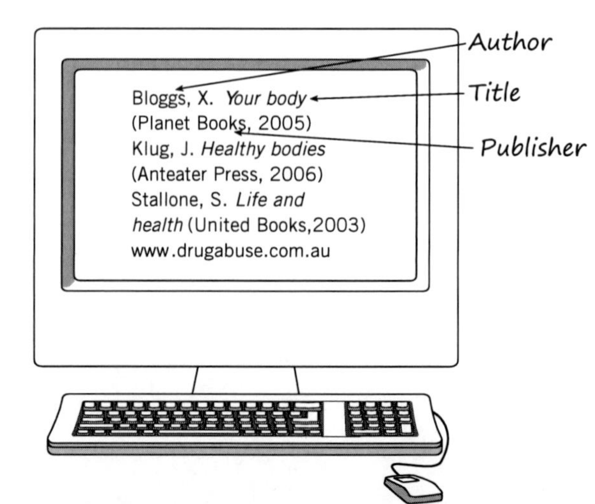

4 **Studying the information**
When you have all the information, that is, when you have filled in all the boxes in the table, the next step is to read it over carefully, trying to understand the links between the causes and effects.

5 **Writing paragraphs**
In *ScienceWorld 8 Workbook* you learned how to write a paragraph. Remember, each paragraph should begin with a topic sentence that states in general terms what the paragraph is all about. The remaining sentences should then give details supporting the topic sentence.

Use *Cause and effect linking words* from page 92 to link the causes and effects for each of the substances.

6 **Writing the first draft**
On the next page is a step-by-step plan for how to structure your essay.

7 **Editing**
Read through your essay. As you read it, ask:
- Does it make sense?
- Does it relate to the task?
- Does it flow well from one point to another?

8 **Estimate the number of words**
(number of lines x words per l ine).
- If there are many more than 300 words, can you write the same information in fewer words without leaving out vital information? Can you omit some details or some examples?
- If there are fewer than 300 words, can you expand on the relevant information with more details or examples?

9 **Proofreading**
Check for errors in spelling, punctuation, grammar and usage.

10 **Write a final draft**, making sure you present it neatly.

Write your final draft using a WORD PROCESSOR program on a computer. These programs not only ensure professional presentation, they also count words, check spelling and sometimes check grammar.

How to structure your essay

Introduction

A general statement about the effects of the three substances on your body previews what you will discuss in the essay.

Alcohol, tobacco and heroin are all drugs that directly affect the body's nervous system.

Body

Paragraph 1

Start with a topic sentence about alcohol. Then describe the effects of alcohol on the nervous system of the body. Use some *Cause and effect linking words.*

Paragraph 2

Topic sentence about tobacco. Then describe the effects of tobacco on the nervous system of the body. Use some *Cause and effect linking words.*

Paragraph 3

Topic sentence about heroin. Then describe the effects of heroin on the nervous system of the body. Use some *Cause and effect linking words.*

Conclusion

Sum up what the three substances have in common and how they affect the nervous system. Do not introduce any new information.

Don't forget to attach your bibliography.

Chapter 8
Using electricity

Overview

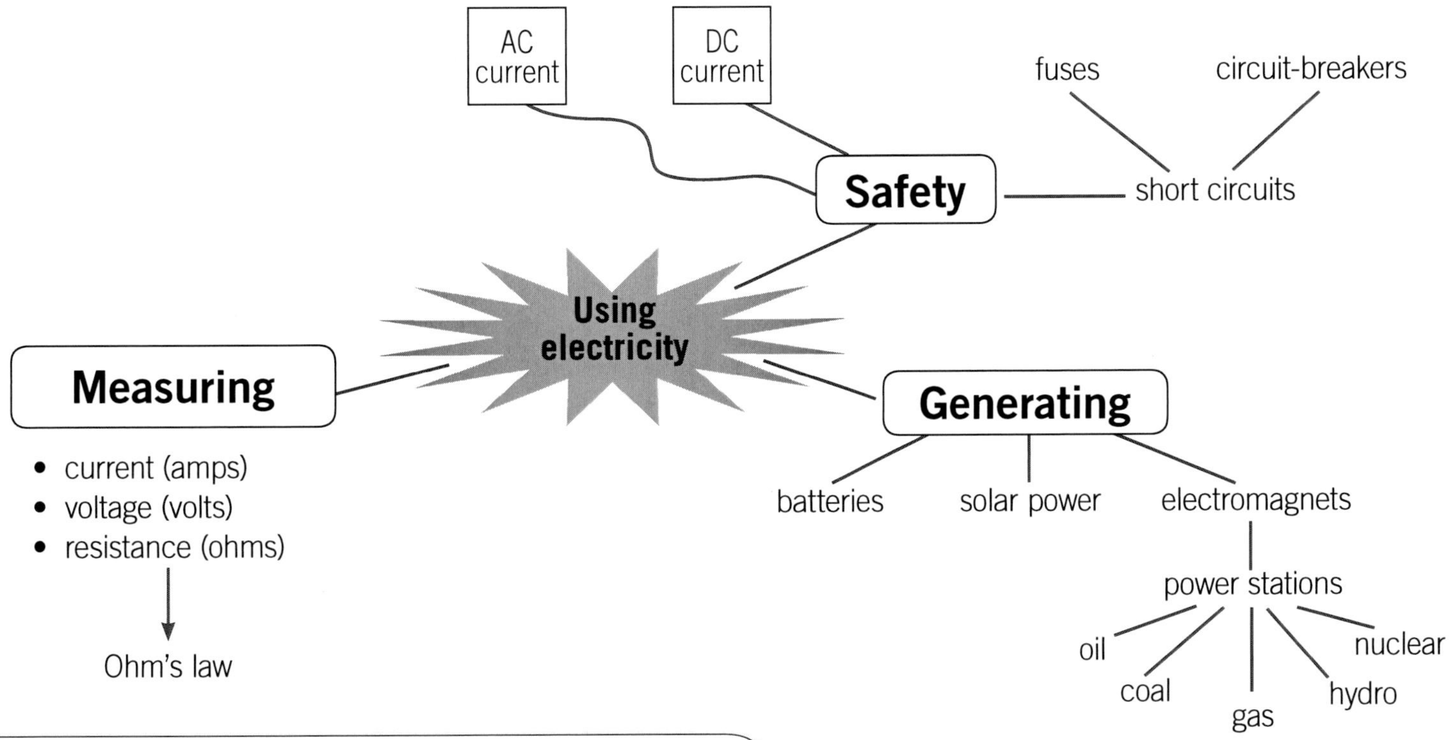

Using electricity

Safety
- AC current
- DC current
- short circuits
  - fuses
  - circuit-breakers

Measuring
- current (amps)
- voltage (volts)
- resistance (ohms)

→ Ohm's law

Generating
- batteries
- solar power
- electromagnets
  - power stations
    - oil
    - coal
    - gas
    - hydro
    - nuclear

What do you know already?

Exercise 1 Explaining

When one light blows in a string of Christmas tree lights, they all go out. When one light blows in a home, the others stay on. Use simple diagrams to explain why this happens.

Christmas tree lights

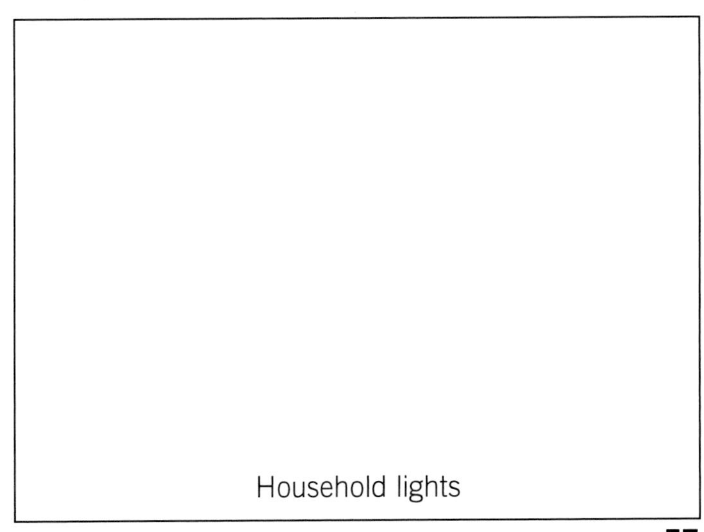

Household lights

During reading

In this chapter you will find Core and Challenge exercises. After reading each section of the textbook, you will find an exercise to complete. If your solution to this exercise is correct (check your answers on page 64), go straight to the Challenge exercise. If you cannot do the exercise, or get it wrong, read the textbook section again or ask your teacher for assistance, then go on to the Core exercise. Of course you can do both the Core and Challenge if you like.

Exercise 2 Interpreting symbols

After reading AC and DC on page 191

Read the information opposite. It is from the information plate on a clock-radio.

1 Is the power source **a** batteries, **b** mains supply, or **c** both?

____

2 What does 9 V DC stand for?

____

3 At what rate does the AC current change direction?

____

AUDIOSONIC MODEL CR-329

230/240 V AC ∿50 Hz
9 V DC Battery back up 6 W

CAUTION: To reduce the risk of electric shock, do not disassemble the product. No user serviceable parts inside. Refer all servicing to an authorised service facility.

MADE IN CHINA Serial no. W234XP112748

Check your answers on page 64 of this workbook.

Core exercise

BREVILLE
MODEL HD-261
240 V ∿50 Hz
1200 W
12 ED 45089
Serial no. Q 95021

1 Is the power supply from (a) batteries,

(b) mains supply, or (c) both? ______________

2 Does the electric current move in the wires in one direction only or in one direction and then in the opposite direction?

____

3 Suggest which appliance this information plate is on?

____

4 What does the diagram on the information plate mean to tell you? ______________

Challenge exercise

(These questions refer to the two information plates on this page.)

1 What do you think this symbol might represent?

____

2 What do 6 W and 1200 W on the information plates stand for?

____

____

____

3 Why would the clock-radio use only 6 W and the other appliance use 1200 W?

____

____

____

____

Exercise 3 Applying information

After reading Fuses and circuit-breakers on page 194 and Earthing on page 195

Study the meter box on the right then answer the questions.

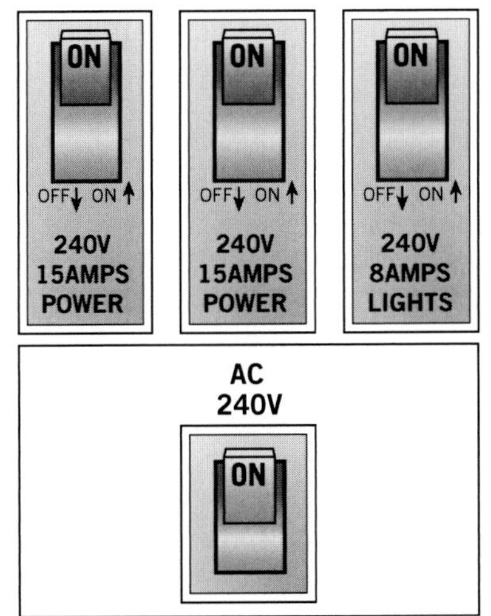

1 How many amps of current can be used for lights before the circuit-breaker switches off?

____

____

2 How are the circuit-breakers for the power circuits different from the one for the lights circuit?

____

____

____

Check your answers on page 64 of this workbook.

Core exercise

1 Below is a circuit containing an 8 amp fuse. If the current reaches more than 8 amps, what will happen to—

a the fuse? ______________________________

b the battery? ______________________________

c the light bulb? ______________________________

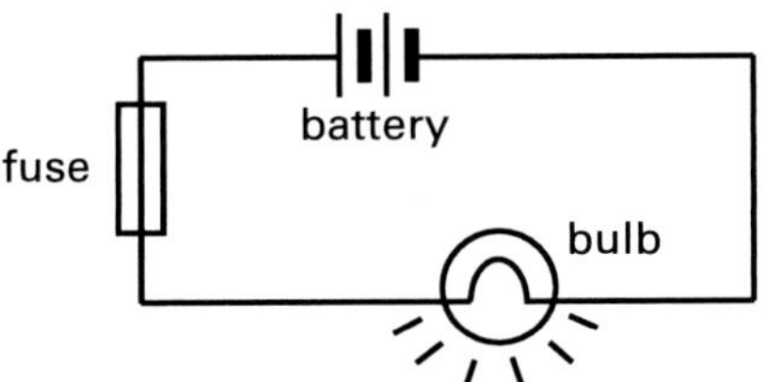

2 In a 3-pin household plug—

a Which colour is the active wire? ______________

The neutral wire? ______________

The earth wire? ______________________

b Which pin is the longest? ______________________

c What is the voltage difference between the active and neutral wires?

____

d What does the earth wire do?

____

Challenge exercise

In your own words, explain how a short circuit occurs. Use a diagram to illustrate your answer.

Exercise 4 Using mathematical formulas

After reading Measuring electricity on pages 198–201

1 Complete the following table.

Name	Symbol	Unit of measurement	Measurement instrument
electric current			
electrical 'pressure'			
resistance			voltmeter and ammeter

2 Use the formula for Ohm's law to calculate the current if the voltage is 24 volts and the resistance is 8 ohms. (Make sure your answer has the correct units.)

Check your answers on page 64 of this workbook.

Core exercise

Use Ohm's law to work out the following:

1 What is the voltage when a current of 2 amps flows through a 40 ohm resistance?

2 How much current flows when the voltage is 12 volts and the resistance is 60 ohms?

3 What is the resistance when the voltage is 240 volts and the current is 15 amps?

Challenge exercise

1 Volts measure *potential difference*. This is the difference between which two things?

2 Why is a voltmeter connected in parallel with the part of the circuit you are measuring, whereas an ammeter is connected in series in the circuit?

3 Is this graph true to Ohm's law? Explain.

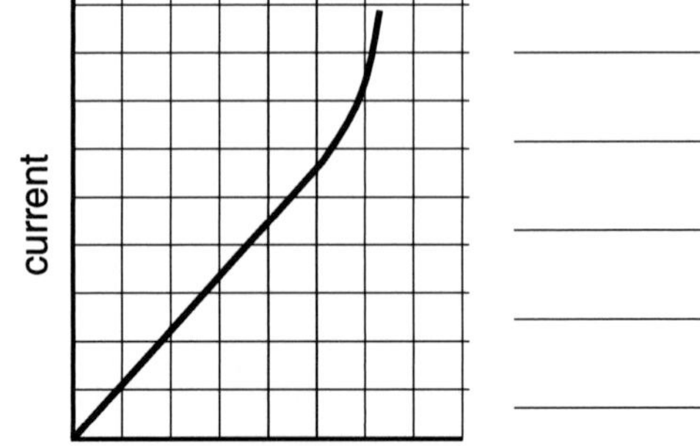

Exercise 5 Applying mathematics to everyday situations

After reading Paying for electricity on page 202

The Smiths have just received their electricity bill.

Meter number	Readings 22/07/06	21/04/06	Tariff	kWh	Months	Amount
Domestic 958296	16260	14057	11	2203	3	228.07
			AMOUNT DUE	05/08/11		228.05

1 What do the letters kWh on the bill stand for? ____________________

2 What was the meter reading on 21 April? ____________________

3 What was the meter reading on 22 July? ____________________

4 How much electricity did the Smiths use over those 3 months? ____________________

Check your answers on page 64 of this workbook.

Core exercise

Below are two meter readings.

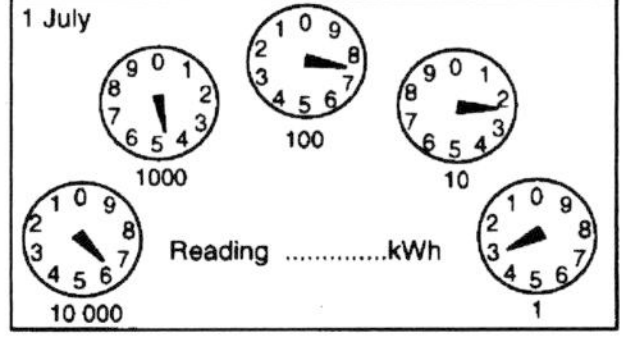

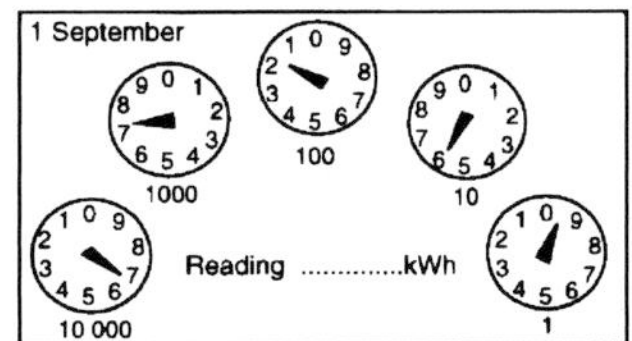

1 Calculate the kWh used between 1 July and 1 September.

____

2 Calculate the electricity bill if—
the first 300 kWh costs 17 cents per kWh,
the next 900 kWh costs 12 cents per kWh,
the rest costs 10 cents per kWh.

Show your working here, or in your notebook if you need more space.

Challenge exercise

Show your working for 3, 4 & 5 in your notebook.

1 If a colour television uses 350 joules per second, what is its power in watts?

____

2 What is another name for 1 000 000 joules per second?

____

3 If $\text{Power (watts)} = \dfrac{\text{Energy produced (joules)}}{\text{time (seconds)}}$

calculate the wattage of a light bulb which uses 600 joules of energy every 8 seconds.

4 Which uses more energy:
10 x 100 watt light bulbs for 5 hours or a 1200 watt electric iron used for 2 hours?

5 Use the table on page 202 of your textbook to calculate how many hours you could use a 75 watt light bulb for the same cost as running a TV set for 3 hours.

Exercise 6 Explaining in your own words

After reading Where does electricity come from? on pages 207–208

Complete the following table.

Method	How it generates electricity	Everday usage
Using ________	The zinc case (–) and the paste inside (+) react and electrons flow between them.	
Using ________		Solar car, pocket calculator, remote telephone
Using magnets		

Check your answers on page 64 of this workbook.

Core exercise

Lin Lee does the Activity on page 208 of the textbook. Complete her report by filling in the blanks.

When I push the magnet into the ________ the needle of the galvanometer moves to the right.

This means an electric ________ has been produced.

The harder I push, the ________ the current.

When I pull the magnet out, the needle moves to the

________ .

When the magnet is ________ there is no current. Even if I hold the magnet still and move the coil

____

____

From these observations I can hypothesise that electricity is generated whenever

____

____

____

Challenge exercise

Examine Fig 7 and Fig 8 on page 207 of your textbook showing a solar car and how a solar cell works. List four factors that would affect the distance travelled and the speed at which a solar car could travel.

____

____

____

____

____

____

____

____

____

____

____

____

Exercise 7 Recalling information in diagrams

After reading Power stations on page 209

After studying Fig 10 and Fig 14 close your textbook and, on the diagram above, label the following.

- boiler
- steam
- turbine
- generator
- cooling tower
- step-up transformer
- step-down transformer
- primary coil
- secondary coil

Check your answers on page 64 of this workbook.

Core exercise

1 In a coal-burning power station, what is it that drives the turbines and generator?

____

2 In a hydroelectric power station, what is it that drives the turbines and generator?

____

3 Why is the electricity produced by power stations passed through a step-up transformer?

____

____

____

4 Electricity is stepped down before being used in houses. What does this mean? Why is it done?

____

____

____

Challenge exercise

Electricity can be generated using many different methods. Complete the table below.

Source of power	Advantage	Disadvantage
oil		
coal		
gas		
uranium		
falling water		
wind		
ocean waves		
sun		

Answers

Exercise 2

1 c both
2 9 volts direct current
3 50 Hz (50 times a second)

Exercise 3

1 8 amps
2 Each of the power circuit-breakers will allow 15 amps of current through it before it switches off. However, the lights circuit-breaker will only allow 8 amps through.

Exercise 4

1

Name	Symbol	Unit of measurement	Measurement instrument
electric current	I	ampere	ammeter
electrical 'pressure'	V	volt	voltmeter
resistance	R	ohm	voltmeter and ammeter

2 $V = IR$ $\quad I = \frac{24}{8}$

$I = \frac{V}{R}$

$R = 3$ amps

Exercise 5

1 kilowatt-hours
2 14 057 kWh
3 16 260 kWh
4 2203 kWh

Exercise 6

Method	How it generates electricity	Everyday usage
Using batteries	The zinc case (–) and the paste inside (+) react and electrons flow between them.	Torch, camera, portable radio
Using solar cells	Light energy from sunlight is converted to electrical energy in silicon cells.	Solar car, pocket calculator, remote telephone
Using magnets	A magnet is moved through a coil of wire to produce a current.	Generators in power stations

Exercise 7

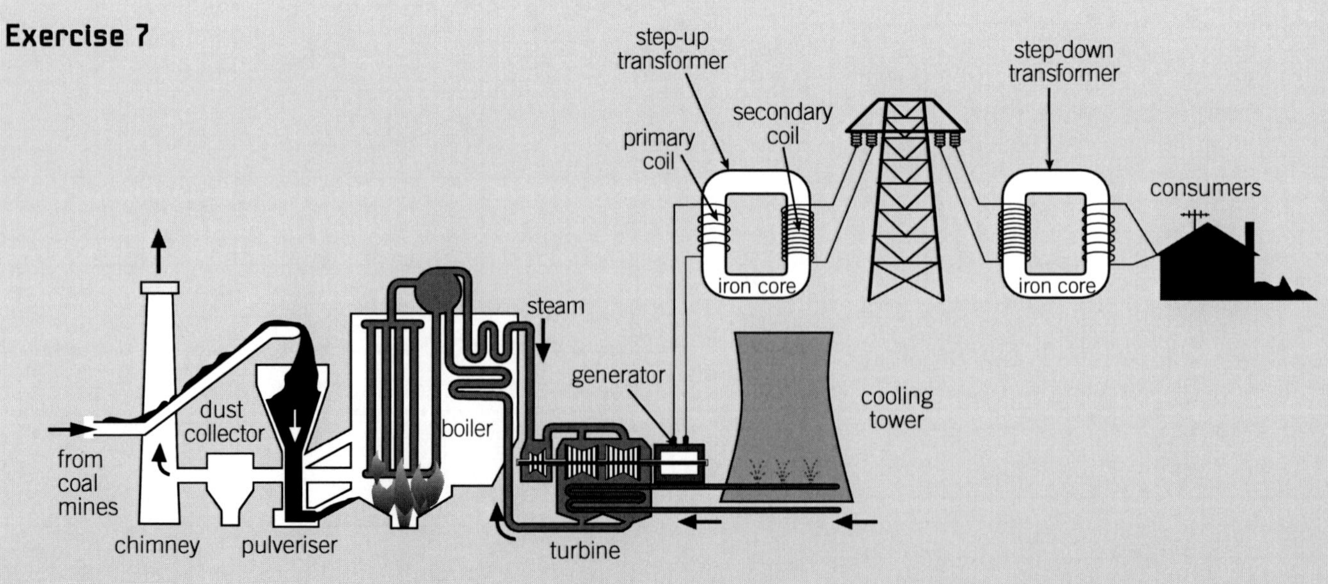

Roundup

Exercise 8 Writing instructions

Electricity is very useful but it can also be very dangerous.

Imagine you have been given the task of designing and writing a leaflet titled *Six ways to prevent electric shock in your home*. The leaflet will be distributed to every household in Australia. Use the information in Chapter 3 to write six major safety rules in the form of instructions. Remember, instructions always start with an action word (verb), e.g. Don't... Buy... Turn... Keep... Put...

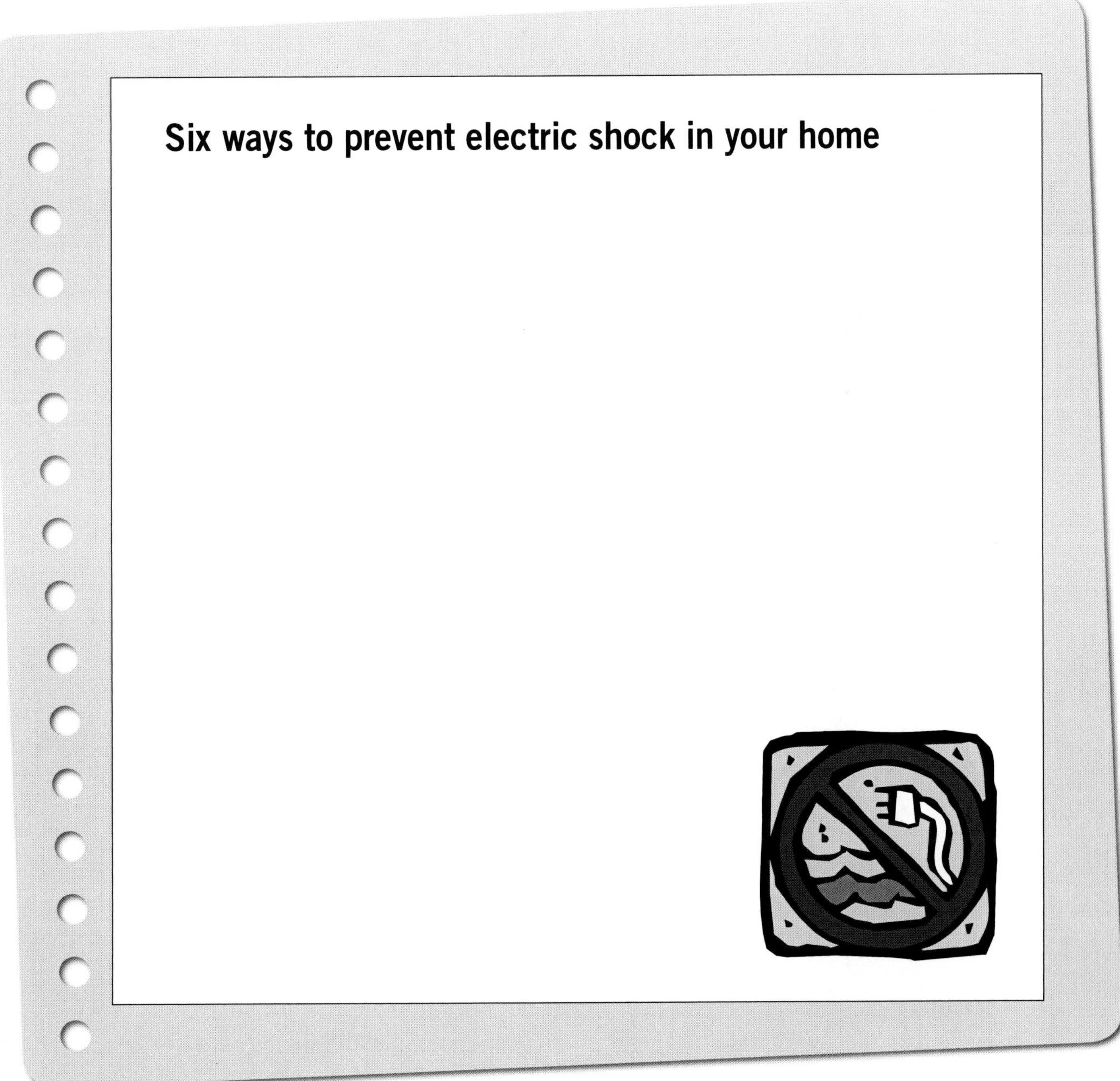

Exercise 9 Reviewing your inferences

Don't forget to go back to Exercise 1 to check whether your explanation was correct.

Chapter 9

Ecosystems

Overview

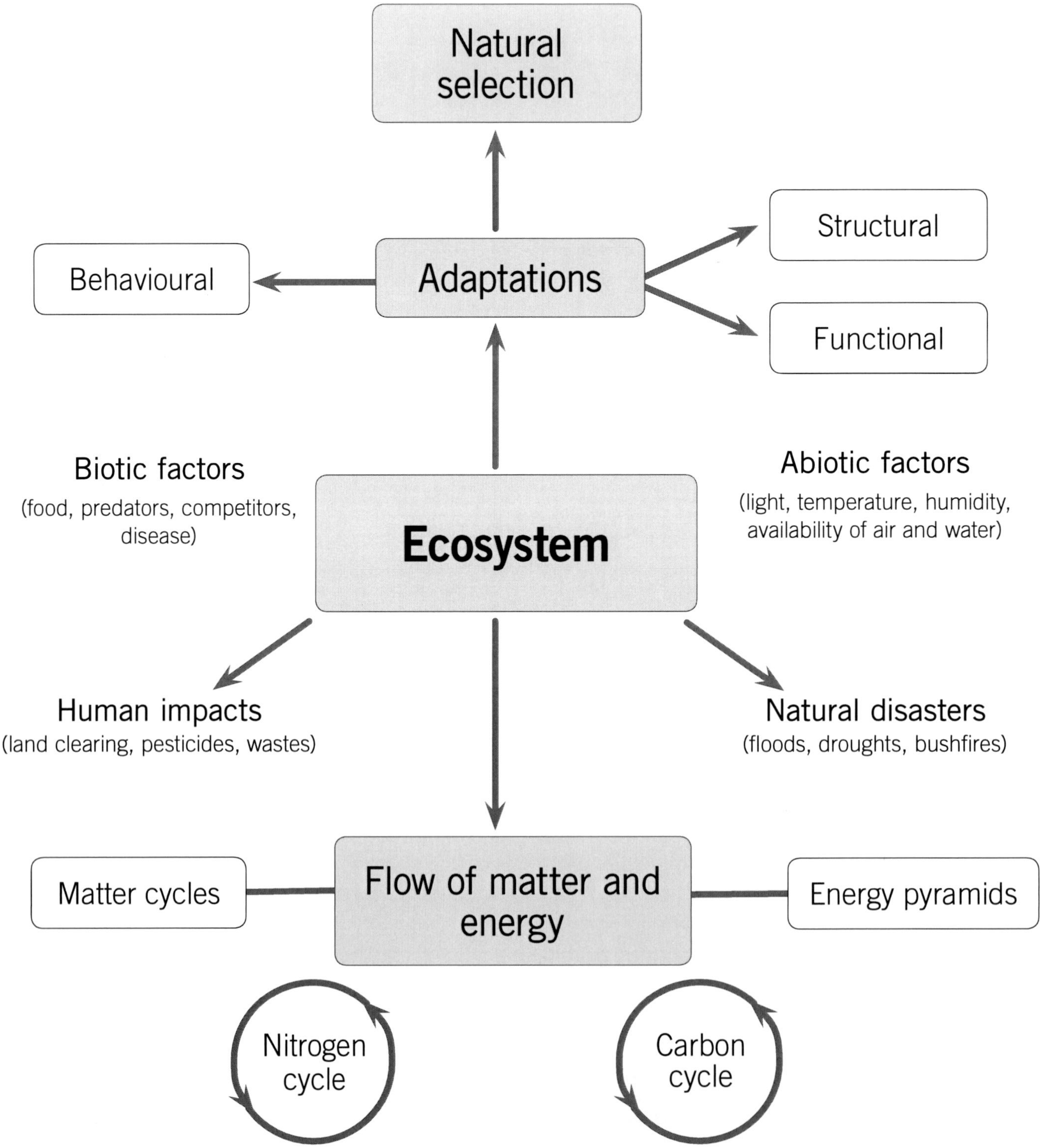

Natural selection

Adaptations

Behavioural

Structural

Functional

Biotic factors
(food, predators, competitors, disease)

Ecosystem

Abiotic factors
(light, temperature, humidity, availability of air and water)

Human impacts
(land clearing, pesticides, wastes)

Natural disasters
(floods, droughts, bushfires)

Flow of matter and energy

Matter cycles

Energy pyramids

Nitrogen cycle

Carbon cycle

What do you know already?

Exercise 1 Explaining

Choose an Australian animal and plant. List three characteristics for each that help it survive in its habitat. You should also explain how each characteristic helps it survive.

Animal: ________________	Plant: ________________
1 ________________	1 ________________
2 ________________	2 ________________
3 ________________	3 ________________

During reading

Exercise 2 Compiling a graphic outline

Knowing how a chapter is organised helps you to understand the relationships among ideas, and it is easier for you to find relevant information quickly.

In this chapter you will be asked questions that require you to find information in the textbook.

To make locating information easier, complete the table below.

Main headings	Sub-headings (List each sub-heading)	Figures (List the figures that have a caption)	Activities and investigations (List the title or describe what it's about, and give the page number)
9.1 Living in ecosystems		Fig 1	Activity page 219: Food webs and the island ecosystem
9.2	Matter and energy in ecosystems		
9.3			

Exercise 3 Organising information in diagrams and tables

After reading Living in ecosystems on pages 219–220

1 Which one of the following diagrams best illustrates the relationship between a population, a community, an ecosystem and an organism?

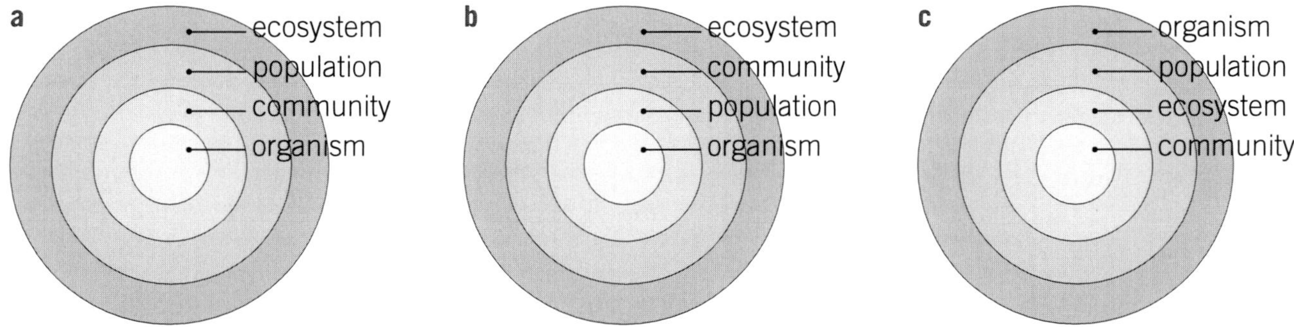

2 Create a table to show whether these factors are BIOTIC or ABIOTIC factors.

- light
- food supply
- temperature
- disease organisms
- competitors
- predators
- humidity
- availability of air, water and nutrients

Working space

Exercise 4 Expressing and justifying points of view

After reading Living in ecosystems on pages 219–220

Here are eight factors that affect the survival of the human species.
Choose the factor that you think is most critical and place a 1 next to it.
For the next most critical factor place a 2 next to it, and so on to the least critical factor.

☐ food ☐ temperature ☐ disease organisms

☐ predators ☐ humidity ☐ availability of air, water and nutrients

☐ competitors ☐ light

When you have finished, discuss your answers with others in the class.

Exercise 5 Classifying

After reading Adaptations on pages 221–222

Arrange the following characteristics of organisms into the correct column in the table below.

- lays hundreds of eggs
- lives in a family grouping
- can change its colour
- has large eyes
- does not sweat
- hibernates in winter
- suckles its young
- appears only at night

Structural adaptations (the shape and size of body parts and how they are put together)	Functional adaptations (how an organism's body works)	Behavioural adaptations (how an organism behaves)

Exercise 6 Designing

Design an animal. This imaginary animal will live in Antarctica. It will eat fish and its only predators are humans. List two structural, two functional, and two behavioural characteristics it will need to survive in this habitat.

Structural adaptations	Functional adaptations	Behavioural adaptations

Exercise 7 Inferring and predicting

After doing Investigation 21 on pages 222–223 and reading pages 222–224

Here is a data table showing the relationship between skin colour and survival rate in a population of a particular type of organism.

Skin colour	Survival rate (per 100 organisms)
black	10%
white	70%
black & white stripes	30%

1 What inference can you make about the habitat in which this population of organisms lives?

2 What prediction can you make about the next generation of this organism?

Exercise 8 Designing an experiment

After doing Investigation 21 on pages 222–223 and reading pages 222–224

Soils in Australia vary in acidity from place to place. Some soils are very acidic (about 5.0 on the pH scale), some are neutral (pH 7.0), while others are alkaline (pH 8.0). One factor in the success of the growth of native plants is their ability to grow in a wide range of soil types.

Suppose you have seeds from a eucalyptus tree, a wattle tree and bottlebrush tree. You want to find out which type of tree grows best in the widest range of soil pHs.

Design an experiment to investigate this. Draw a labelled diagram of your set-up.

Diagram of set-up

Exercise 9 Answering multiple-choice questions

In Chapter 2 of this workbook, you learned how to answer and write multiple-choice questions.

This is a very common method of testing your understanding in science, so the following exercises will give you lots of practice!

Stop at the end of each question to check your answers on page 75. Do not continue until you know exactly why any of your answers were incorrect by re-reading the textbook or checking with your teacher.

Use a pencil to shade the circled letter next to the answer you think correctly completes the statement. For example—

Insects have (a) four legs
(b) six legs
(c) eight legs

After reading Matter and energy in ecosystems on pages 227–232

1 In an ecosystem—
(a) all matter is passed from one organism to another through a food chain
(b) only a small percentage of energy is recycled
(c) there are equal numbers of producers and consumers

2 Food and water consumed by an organism are used for—
(a) growth of new tissues
(b) respiration
(c) both (a) and (b)

3 As you go up the food web from producers to higher order consumers, the mass of food—
(a) increases
(b) decreases
(c) stays the same

4 Decomposers are—
(a) microscopic organisms that cause dead animals to decay
(b) animals that eat the flesh and organs of dead animals
(c) fungi

5 Carbon is recycled into the air when—
(a) coal, oil and wood are burnt
(b) decomposers break down the bodies of dead organisms into carbon dioxide
(c) both (a) and (b) occur

6 Nitrogen from plants and animals is recycled into the air through—
(a) the breakdown of protein in cells
(b) the action of bacteria in the soil
(c) lightning strikes

7 Mass is lost from a food web in the form of—

(a) heat

(b) wastes

(c) wastes and products of respiration

8 As you go up the energy pyramid from producers to higher-order consumers, the amount of energy—

(a) increases

(b) decreases

(c) stays the same

9 A diet consisting mainly of plants and plant products (seeds, nuts, oils) is—

(a) more energy efficient than a mainly meat diet

(b) less energy efficient than a mainly meat diet

(c) the same in energy efficiency as a mainly meat diet

10 Survival of an organism depends on—

(a) biotic (biological) factors

(b) abiotic (physical) factors

(c) both biotic and abiotic factors

11 Availability of food is—

(a) a biotic factor

(b) an abiotic factor

(c) both a biotic and an abiotic factor

12 Availability of clean water and air is—

(a) a biotic factor

(b) an abiotic factor

(c) both a biotic and an abiotic factor

13 The nitrogen in the bodies of animals comes from—

(a) nitrogen in the air

(b) nitrates in the soil

(c) nitrogen in the proteins of other organisms

14 The energy contained in the organisms in all levels of a food web originally comes from—

(a) nutrients in the soil

(b) sunlight

(c) other animals

15 Fourth-order consumers are rare in a food web because—

(a) the animals would have to be very large

(b) the loss of energy through the food web cannot usually support these consumers

(c) the third-order consumers release too much energy

Check your answers on page 75 of this workbook.

Exercise 10 Completing diagrams

1 Complete the carbon cycle.

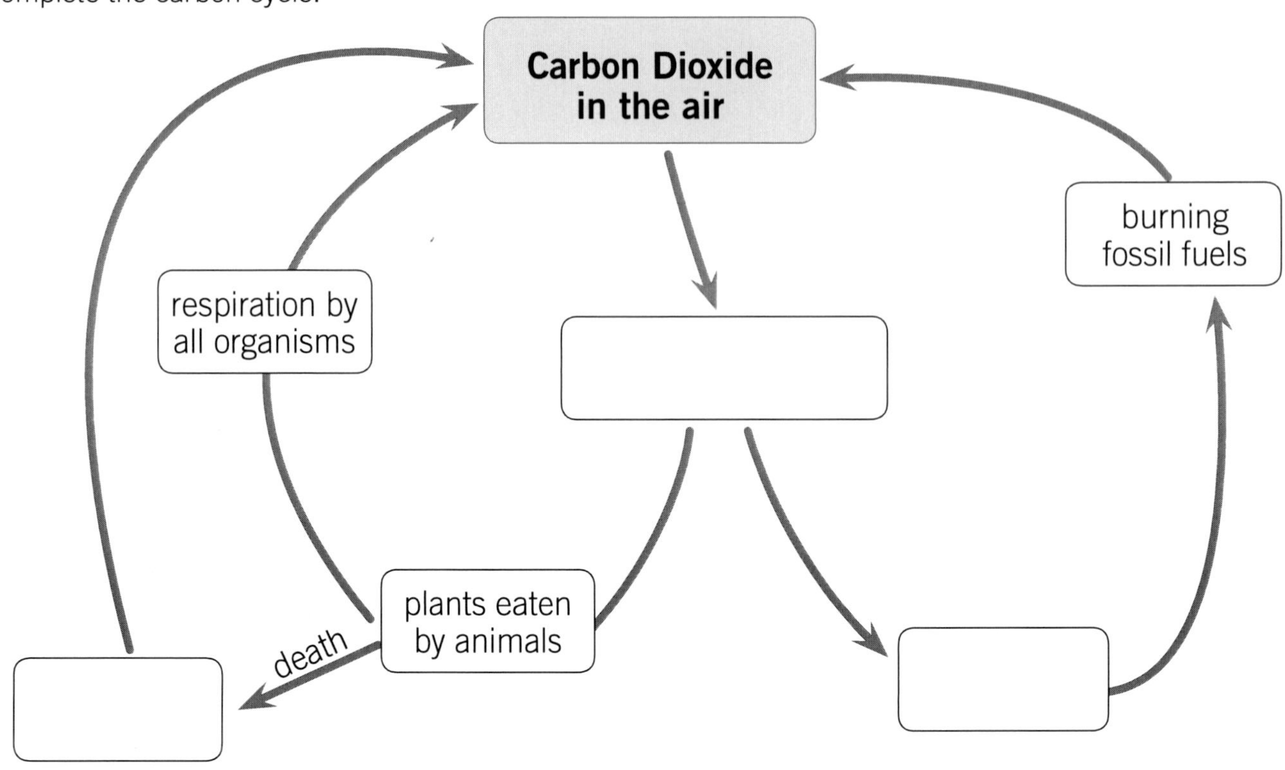

2 Complete the nitrogen cycle.

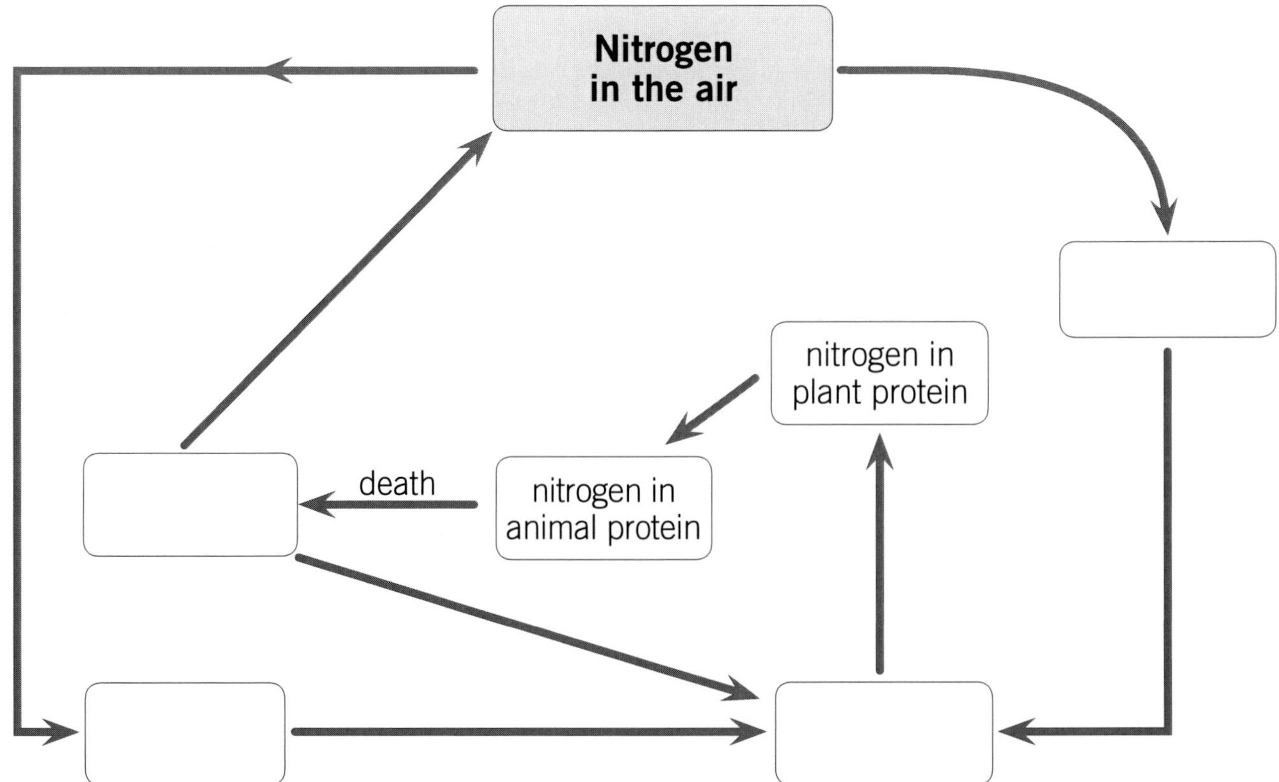

Exercise 11 Recalling words and meanings

In this chapter, there were many scientific words for you to learn and to understand. Complete the table by adding the missing letters in the right-hand column or writing the meanings in the left-hand column.

Meaning	Word or words
1 an organism that eats another organism	c _ n _ _ _ _ r
2 organisms that decay dead organisms	_ _ co _ _ _ s _ _ s
3 a process in living things that produces energy	r _ _ _ _ _ a _ _ _ n
4 subst ances containing nitrogen, found in organisms	_ _ o _ _ i _ s
5	ecosystem
6	biotic factor
7	predator
8 a chemical that is designed to kill organisms, particularly insects	_ e _ _ i _ _ d _
9 a type of adaptation that describes the working of an organism's body	_ _ _ _ t _ _ n _ _
10 describes the types of characteristics that increase an organisms chances of survival in a particular habitat	_ a _ _ _ _ a _ _ e

Answers

Exercise 9

1 b	6 b	11 a
2 c	7 c	12 b
3 b	8 b	13 c
4 a	9 a	14 b
5 c	10 c	15 b

Roundup

Exercise 12 Recalling, discussing and applying

Work in a group to respond to the following task. Elect a recorder who will take notes of the discussion and write down the main points of the discussion. Each member of the group will then take a copy of the notes and, individually, complete the task.

There are five groups of vertebrates in the animal kingdom— fish, amphibians, reptiles, birds and mammals. Animals in only three of these groups live successfully in the polar regions of the Earth (the Arctic and Antarctica). Give detailed descriptions of the adaptations of the animals in these groups that enable them to live in these habitats.

Animals in three groups of vertebrates have adapted to live successfully in the polar regions of the Earth.

These groups are ______________________________

Chapter 10

Dynamic Earth

Overview

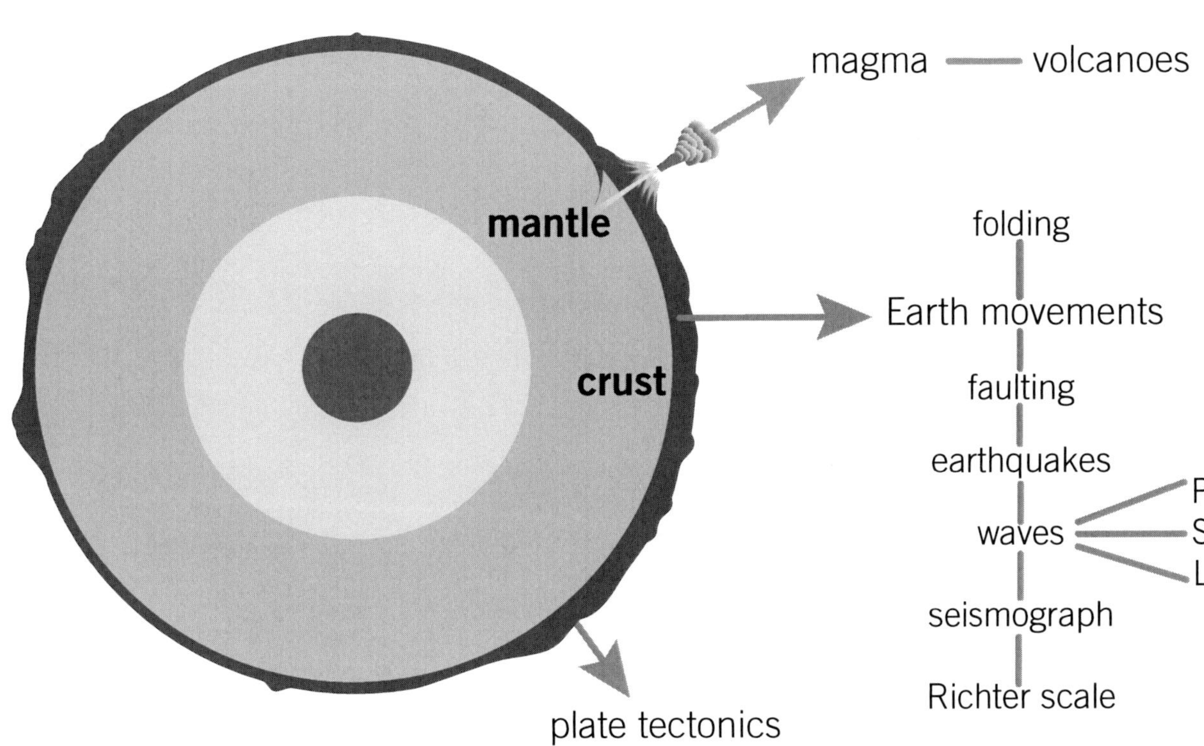

What do you know already?

Exercise 1 Inferring

List two possible inferences you can logically make based on the evidence below.

Evidence: Fossilised marine shells have been found in the rocks on Mt Everest.

1 ____

____

2 ____

____

During reading

Exercise 2 Recalling information

After reading Earth's movements on pages 246–248

Match each word with its meaning.

Words	Meanings
dormant •	• innermost part of the Earth made of a liquid outer section and a solid inner section
molten •	• a downwards-bending fold in rock layers
crust •	• buckling of rocks caused by huge Earth forces
mantle •	• an upwards-bending fold in rock layers
core •	• liquified by extreme heat
magma •	• relatively thin, solid outer layer of the Earth
lava •	• the mouth of a volcano
folding •	• a crack in the Earth's surface along which rocks move
fault •	• thick layer of rock below the Earth's crust; it is partly solid and partly molten
syncline •	• thick, hot, molten rock on the Earth's surface
anticline •	• inactive
crater •	• thick, hot, molten rock under the Earth's surface

Exercise 3 Predicting using models

After reading Earth's movements on pages 246–248

1 Imagine you have three layers of plasticine stuck together and two flat bricks at each end.

You push the bricks towards each other with equal force. Predict the shape of the plasticine and sketch it in your notebook, showing the three layers.

2 Now imagine you have three layers of plasticine stuck together and two flat bricks at each end. This time you cut the plasticine right through along the dotted line.

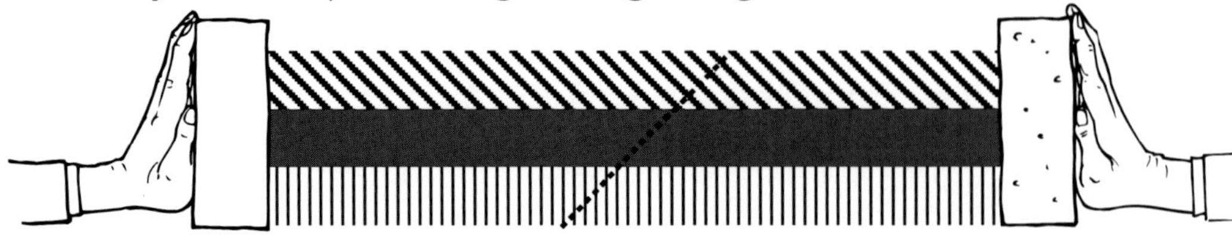

You push the bricks towards each other with equal force. Predict the shape of the plasticine and sketch it in your notebook, showing the three layers.

Exercise 4 Recalling information

After reading Earthquakes on pages 252–254

Match each word with its meaning.

Words	Meanings
earthquake •	• an instrument for detecting and measuring earthquakes
shock waves •	• waves which result from the release of stored-up energy in rocks
epicentre •	• also called compression waves; the first waves recorded on a seismogram
focus of earthquake •	• series of vibrations at the Earth's surface caused by movement along a fault line
vibration •	• also called shear waves; the second waves recorded on a seismogram
seismograph •	• the point on the Earth's surface directly above the focus of an earthquake
primary waves (P) •	• the most damaging waves in an earthquake and the last recorded on a seismogram
secondary waves (S) •	• the place under the Earth's surface where the rocks move
surface waves (L) •	• a rapid to and fro movement

Exercise 5 Interpreting a graph

After reading pages 254–256

Examine the graph on page 254 very carefully.

1 Calculate the distance from the earthquake if—

  a the difference between the arrival times of the P and S waves is 100 s. ____________

  b the P waves arrive at 8.00 am and the S waves arrive at 8.03 am. ____________

  c the P waves arrive at 11.22.30 and the S waves arrive at 11.27.15. ____________

2 What is the difference in times of arrival of the P and S waves if—

  a the earthquake is 4000 km away? ____________

  b the earthquake is 2800 km away? ____________

3 Write the exact time that the S waves will arrive from an earthquake 3000 km away if the P waves arrive at 1.15.30. ____________

Exercise 6 Interpreting information

After reading The Richter scale on page 257

1 Why do scientists need a Richter scale? ______________________________

____

2 Study the bottom photo on page 252 of your textbook. Use the table on page 257 to decide which magnitude on the Richter scale this earthquake might have been. Give a reason for your answer.

Magnitude = ____________

Reason— ______________________________

____

____

3 If a magnitude 5 earthquake releases as much energy as 20 000 tonnes of TNT,
how many tonnes of TNT would be the equivalent of—

a a magnitude 4 earthquake? ____________ **b** a magnitude 8 earthquake? ____________

Exercise 7 Writing in role

After reading page 257

Complete this eyewitness account from a survivor of an earthquake.

Rema, aged 42, mother of three young children – earthquake magnitude 5.5

I was awoken by my children crying ______________________________

____

____

____

____

____

Billy, aged 13 – earthquake magnitude 8.3

It was the most terrifying experience of my life. I am very lucky to be alive.

Firstly ______________________________

____

____

____

____

Exercise 8 Writing a newspaper article

After reading Earthquakes on pages 252–257

The purpose of a newspaper article is to inform. The audience is the general public. A newspaper article is usually written with the most important information first, tapering off to the least important information at the end.

1 Below are two columns. In the first, the structure and features of a newspaper article are described. In the second, a model article has been written for you.

Your task is to match the structure and features with the model article, using arrows.

Were any of the features not included in the article?

Headline

- attention grabbing
- secondary headline may be added in smaller print

Lead paragraph

- tells four Ws—who, what, when and where in 20–30 words
- sometimes tells how and why

Body paragraphs

- most important and interesting information first, least important last
- short paragraphs – about 30 words in each
- each paragraph independent of other paragraphs – no need to link

Language features

- plain, brief, factual language
- adjectives and adverbs kept to a minimum
- quotes of key people involved are used to add interest

Format

- pictures can be incorporated to add interest
- set out in columns

Earthquake Terror

Thousands flee city

An earthquake, measuring 6.5 on the Richter scale, struck Mexico City at 3.24 am yesterday as residents slept. The toll from this massive quake currently stands at 124 dead and hundreds more injured.

Thousands of residents, fearing further tremors, are fleeing the city in panic. Huge sections of the city were evacuated when it was feared aftershocks could cause more damage to already weakened buildings.

Emergency workers fear that there are hundreds of people trapped in the rubble. "We can hear their desperate cries for help," a spokesman said. Heat-seeking and listening devices are being used to locate victims.

The quake's epicentre was recorded only 6 km west of Mexico City, and lasted 22 seconds.

The damage bill is expected to be more than $4 billion.

2 Now it's your turn to be the reporter. Research one of the earthquakes listed on the right. Using the structure and features of a newspaper article described above, write your own newspaper article in your notebook. Write in two columns. You may choose to add diagrams and pictures to add interest.

Shaanxi, China on 24 January 1556
San Francisco, USA on 18–19 April 1906
Yokohama, Japan on 1 September 1923
Mexico City, Mexico on 19 and 21 September 1985
Newcastle, Australia on 28 December 1989
Northwest Turkey on 17 August 1999
Java, Indonesia on 27 May 2006
Maule, Chile on 27 February 2010

Exercise 9 Using evidence and applying information

After reading Earth plates on pages 259–261

1 **a** In your own words, explain Wegener's continental drift hypothesis.

b Which three pieces of evidence support this hypothesis?

c Suggest another possible explanation for this evidence.

2 **a** In your own words, explain the theory of plate tectonics.

b Which two pieces of evidence support this hypothesis?

c Suggest another possible explanation for this evidence.

3 In your own words, explain why New Zealand has so many earthquakes and volcanoes and Australia does not.

4 If the theory of plate tectonics is correct, what might happen to New Zealand in millions of years' time?

Roundup

Exercise 10 Completing an overview

Use these words to complete this overview.

- epicentre
- magma
- plate tectonics
- inner core
- crust
- faulting
- folding
- seismographs
- outer core
- earthquake
- volcano
- mantle
- primary waves
- secondary waves
- surface waves
- Richter scale

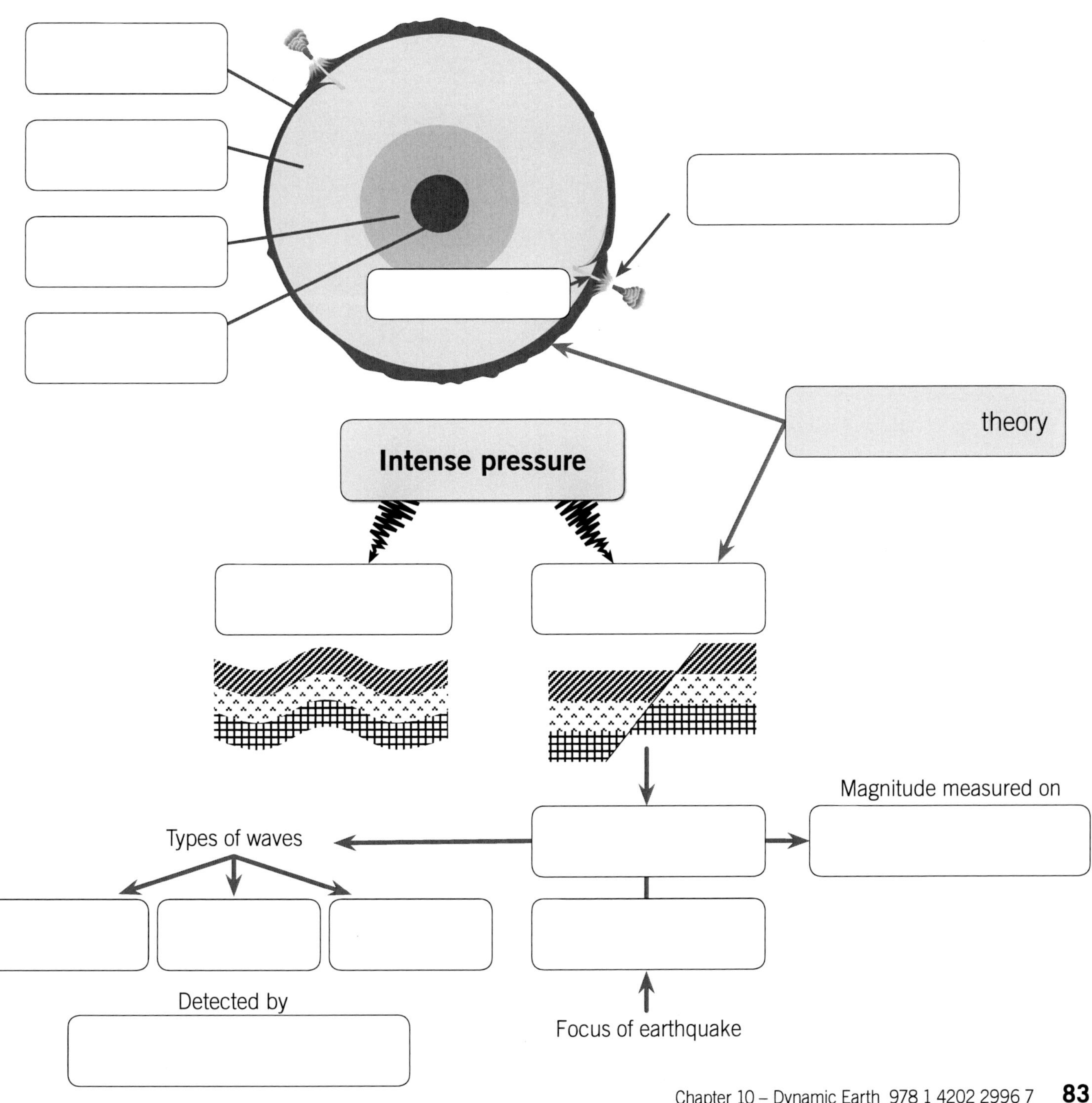

Chapter 11
Communication technology

Overview

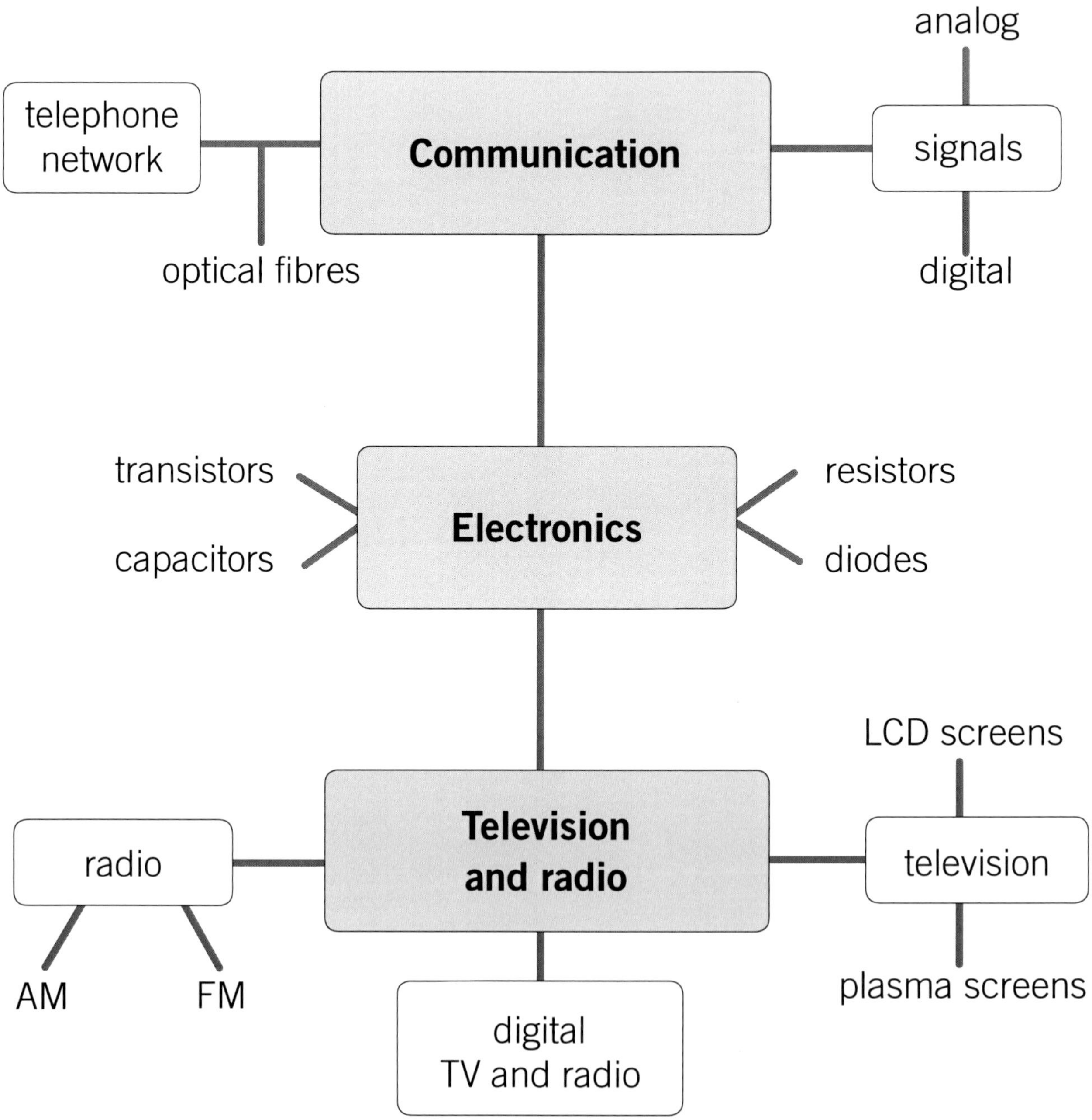

What do you know already?

Exercise 1 Using context clues

In this chapter, you will need to understand the meanings of the words in *italics* in the sentences in the first column of the table below.

Before you look up the meaning in the Glossary section of the textbook or in a dictionary, write in your own words what you think the meaning might be. Use the clues from the other words in the sentence to try to work out a meaning that makes sense.

Word in context	Possible meaning	Glossary or dictionary meaning
1 The greater the *amplitude* of the radio wave, the louder the signal.		
2 When the spies intercepted the message, they had to *decode* it before they could make any sense out of it.		
3 My computer screen has 1 million *pixels* and the image is very clear.		
4 FM 105.4 operates at a *frequency* of 105.4 million cycles per second.		
5 Before the 1980s, all electrical *transmissions* between telephone exchanges were analog.		
6 The use of *integrated* circuits has reduced the size of electronic devices such as computers and mobile phones.		

During reading

Exercise 2 Locating information

In this chapter you will be asked questions that require you to search for information throughout the *whole* chapter in your textbook. To answer a question you should:

1 **Scan** the whole chapter and make yourself familiar with how the authors have organised the material.

2 **Read** the headings and sub-headings. These are 'signposts' to the answers you need.

Complete the exercise below, in which you are given a list of the headings and sub-headings used in Chapter 10.

Read the list of topics and decide which of the headings or sub-headings would be the best place to locate information on each topic. Write in the heading or sub-heading where you would expect to find the information.

Main headings and sub-headings	Topics	I would expect to find information on the Topic under the sub-heading ...
11.1 Communication • Communication devices • Digital and analog signals • The telephone network • Mobile phones • Faxes and modems • Optical fibres **11.2 Electronics** • Semiconductors **11.3 Television and radio** • How does television work? • Digital TV • LCD screens • Plasma screens • Radio and TV transmission • Radio and TV reception • Digital radio	**a** Antennas **b** The advantages of fibre optic cables **c** Electron guns in picture tubes **d** How to connect a diode properly into an electric circuit **e** Microphones **f** Resistors **g** Symbols for diodes and transistors **h** The difference between AM and FM radio **i** The difference between analog and digital signals **j** How sound waves are changed and transmitted by telephones **k** Liquid crystal displays	

Exercise 3 Understanding the information

Read the two paragraphs on the right carefully, then answer the questions below.

> Throughout today you have been the sender and receiver of information. Most of this communication has been transmitted over a short range—by speaking and gesturing with other people close by. However, this short-range communication can be replaced by long-range communication by using various methods such as post, telephone, fax, email and radio.
>
> Communication devices such as a telephone require voice to be changed into electrical signals or light pulses. These signals are transmitted over long distances and then changed back into voice, which is heard by the receiver. Voice is changed into electrical signals by a microphone and the electrical signals are changed back to voice by a speaker.

1 Suggest a heading for the extract—one that sums up the main ideas. ____________________

2 Identify the contrast linking word in the first paragraph. ____________________

3 Which two things are being contrasted in the first paragraph? ____________________

4 Give two examples of short-range communication. ____________________

5 Which words in the first paragraph signal to you that an example or examples will follow? ____________________

6 Identify the sequence linking word in the second paragraph. ____________________

7 Give another word that means the same as transmitted. ____________________

8 Which device is the reverse of a microphone? ____________________

Exercise 4 Working independently

After reading the whole chapter

On the following two pages you will find a circle surrounded by questions.

You are to work independently, at your own rate. You can start answering questions at any point and work your way around until you return to your starting point. Write your answers in the spaces around the questions. If you need more room use an extra sheet of paper.

Try to answer as many questions as possible without using your textbook. If you don't know the answer, use the headings and sub-headings of your text to find the information you need. Then follow these steps:

1 Read the question again.

2 Ask yourself, what exactly is the question asking me to find or do?

3 Read the section very carefully to:

   a understand the main idea

   b identify the supporting details

   c look for linking words that indicate the relationships among ideas. (For example, 'because' and other *Cause and effect linking words* may indicate the answer to a *Why* question.)

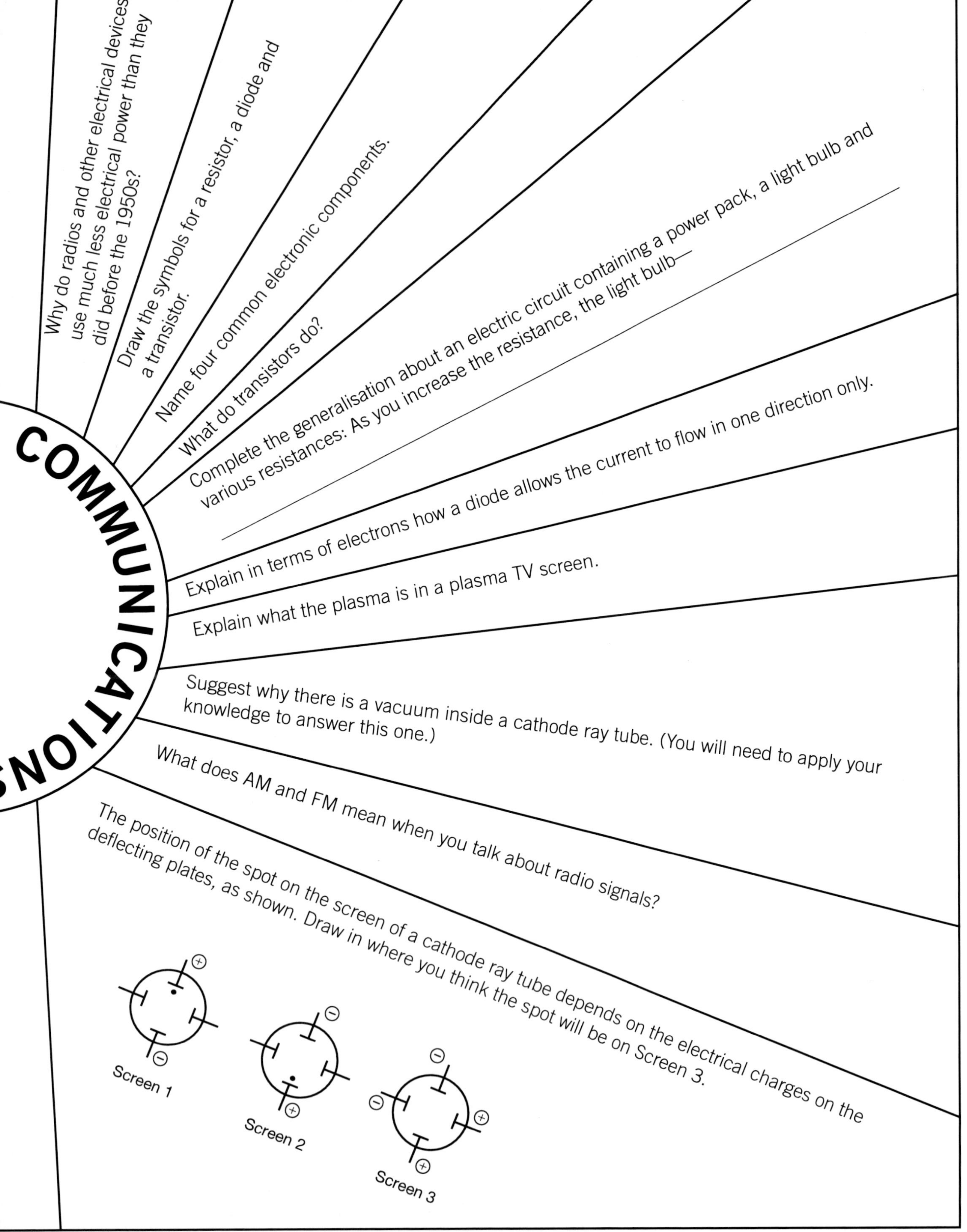

COMMUNICATIONS
Why do radios and other electrical devices use much less electrical power than they did before the 1950s?
Draw the symbols for a resistor, a diode and a transistor.
Name four common electronic components.
What do transistors do?
Complete the generalisation about an electric circuit containing a power pack, a light bulb and various resistances: As you increase the resistance, the light bulb—
Explain in terms of electrons how a diode allows the current to flow in one direction only.
Explain what the plasma is in a plasma TV screen.
Suggest why there is a vacuum inside a cathode ray tube. (You will need to apply your knowledge to answer this one.)
What does AM and FM mean when you talk about radio signals?
The position of the spot on the screen of a cathode ray tube depends on the electrical charges on the deflecting plates, as shown. Draw in where you think the spot will be on Screen 3.
Screen 1
Screen 2
Screen 3

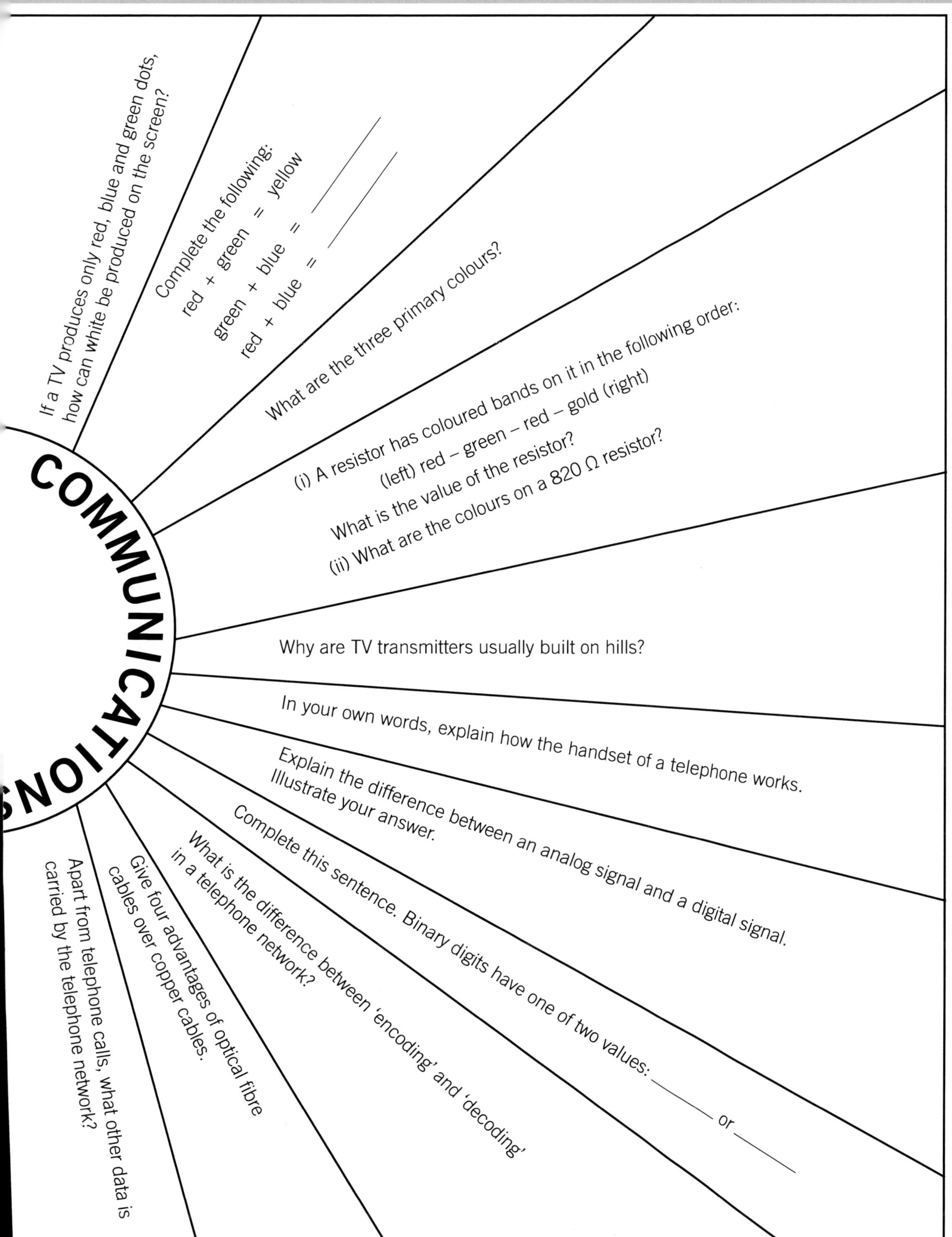
COMMUNICATIONS
If a TV produces only red, blue and green dots, how can white be produced on the screen?
Complete the following:
red + green = yellow
green + blue =
red + blue =
What are the three primary colours?
(i) A resistor has coloured bands on it in the following order:
(left) red – green – red – gold (right)
What is the value of the resistor?
(ii) What are the colours on a 820 Ω resistor?
Why are TV transmitters usually built on hills?
In your own words, explain how the handset of a telephone works.
Explain the difference between an analog signal and a digital signal. Illustrate your answer.
Complete this sentence. Binary digits have one of two values: or
What is the difference between 'encoding' and 'decoding' in a telephone network?
Give four advantages of optical fibre cables over copper cables.
Apart from telephone calls, what other data is carried by the telephone network?

Exercise 5 Writing clues

In the chapters in this book you do crosswords by reading the clues and writing in the appropriate word or words. In this exercise you have to write the clues for the words in the crossword.

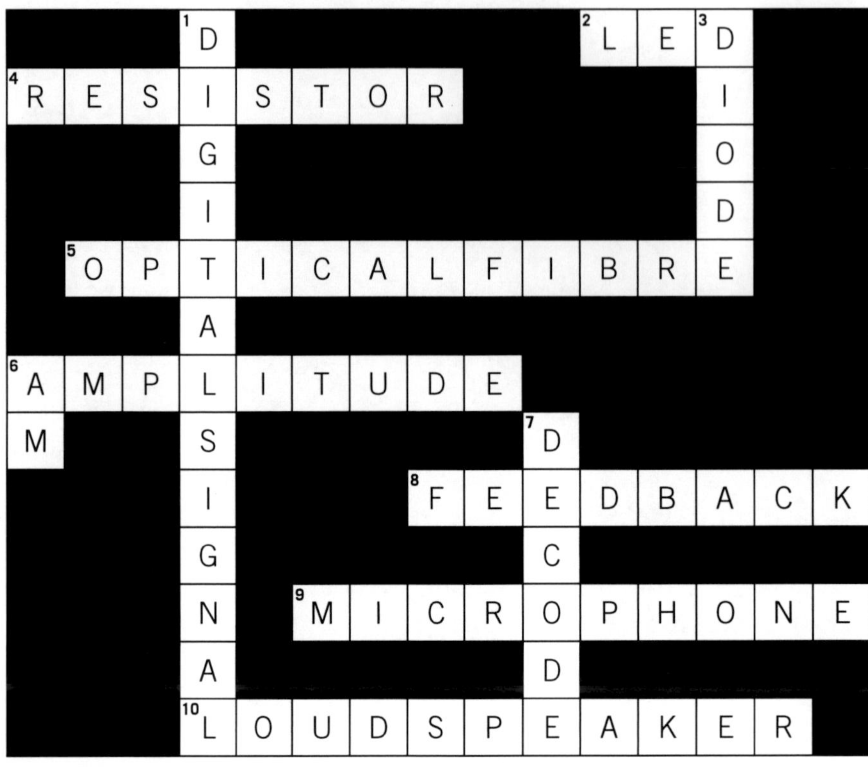

Across clues

2

4

5 (2 words)

6

8

9

10

Down clues

1 (2 words)

3

6

7

Roundup

Exercise 6 Designing a buzzer

Mr Ham, the science teacher, is fed up with students tilting backwards on their chairs. He wants to design an electrical device that sets off a buzzer whenever the front legs of a chair leave the floor.

Use what you know about electric circuits to set up the whole classroom so that Mr Ham can tell at a glance who is tilting back on a chair.

Working space

Linking words

Sequence linking words (for actions that occur one after the other)

first, second	then	before	finally	subsequently
to begin with	later	next	gradually	initially
when	from there	last	initial	after

Cause and effect linking words (for actions which cause another action to occur)

therefore	so that	an outcome of	results in
due to	as a result of	causing	consequently
since	because	caused by	this leads to
the effect of	the reason for	if … then	when … then
resulting in	making	as	accounts for
so	thereby	causes	gives rise to
creates	owing to	hence	thus

Contrast linking words (to describe differences)

on the other hand	in contrast to	different from	however
alternatively	but	although	rather than
whereas	while	unlike	yet
even though	less than	more than	whilst

Comparison linking words (to describe similarities)

both	same	in both	like
in both cases	similarly	and	as well as
in the same way	just as … so …	alike	similar in that

Generalisation linking words (for statements that are true in most cases)

generally	usually	commonly	many
normally	as a general rule	most	mostly